BRETT CARROLL

Chatgpt for Restaurants

60+ Proven Prompts to Streamline Operations, Optimize Daily Workflow, and Accelerate Business Growth

First edition

ISBN: 978-1-7638862-1-6

This book was professionally typeset on Reedsy.
Find out more at reedsy.com

Naam Pla Thai

"Where my love for food and travel truly began. Thank you, Jay and Poh, for your unwavering support, and to the many customers who became dear friends along the way."

Contents

1

Introduction: Transforming Restaurant Management with ChatGPT

For over 10 years, I worked in nearly every role within the restaurant industry, starting at the very bottom and working my way up. My journey began at a bustling Thai restaurant, where I worked as a cleaner, prep cook, and kitchen assistant, gaining firsthand experience in the fundamentals of restaurant operations. From there, I progressed to serving tables and assisting behind the scenes, eventually stepping into management within a few years. Throughout my journey, my passion for Thai food grew, and I gained a thorough understanding of both front-of-house and kitchen operations. I wasn't just a manager — I had worked in every corner of the restaurant, which provided me with a deep understanding of the challenges and intricacies of running a successful food business.

In addition to my restaurant experience, my love for Thai cuisine and Asian culture has shaped much of my life. I have extensively traveled across Asia, particularly Thailand, Japan, Malaysia, and Vietnam, to immerse myself in the cultures and learn more about their rich culinary traditions. From street food stalls to high-end restaurants, I've explored every aspect of the food scene, gaining a deep understanding of flavors, techniques, and ingredients. I love to cook, and my expertise in Thai cuisine goes far beyond just eating it — I enjoy creating dishes and experimenting with traditional recipes. This passion for food, especially Asian cuisine, continues to inspire my work and fuels my desire to share my knowledge with others.

As a restaurant manager, I encountered countless challenges: managing a diverse team, balancing customer expectations, ensuring that operations ran smoothly, and handling the never-ending flow of tasks and communication. One of the most challenging parts of the job was juggling multiple responsibilities—responding to customer inquiries, creating marketing materials, developing menus and profitable 'winning dishes,' and organizing staff schedules. There were always more things to do than hours in the day, and despite my best efforts, it often felt like I was just treading water.

Looking back, I can't help but wish ChatGPT had existed when I was managing the restaurant. This AI-powered tool would have been a game-changer, helping me streamline operations, create marketing materials, respond to customers more efficiently, and save valuable time. Little did I know that my journey in the restaurant world would eventually lead me to the exciting realm of artificial intelligence, where I could better understand how technology could address these common pain points.

In a recent position, I spent twelve months training AI models, working closely with systems in their test stages. This hands-on experience gave me a deep understanding of how AI works and, more importantly, how it can be used in real-world applications. It was during this time that I discovered the potential of ChatGPT — an AI model capable of generating human-like text and completing tasks that once required hours of effort. It wasn't long before I saw how transformative ChatGPT could be for the restaurant industry. From automating marketing tasks to handling customer service inquiries and improving operational processes, I knew that this tool could help restaurants run more efficiently than ever before. Additionally, by streamlining workflows and reducing overhead costs, it has the potential to boost profitability and enhance budget management.

This book is the culmination of my years of experience in the restaurant business and my recent exploration into the world of AI. I've created it to help restaurant owners and managers just like you save time, improve customer engagement, enhance the efficiency of your operations, and ultimately boost profitability. Inside, you'll find over sixty proven ChatGPT prompts specifically designed for the restaurant industry. These prompts are not theoretical; they've been tested by my former colleagues, current restaurant managers, and other industry professionals throughout the creation of this book. Whether you're looking to improve your social media marketing, streamline internal communications, or craft personalized responses to customers, these actionable prompts will help you get the most out of ChatGPT.

As someone who has spent years in the trenches of the restaurant world, I understand the pressure and challenges that come with managing a restaurant. Now, with the help of ChatGPT, I want to show you how AI can make your job easier and more rewarding. Together, let's dive into the future of restaurant management, where technology can help you work smarter, not harder.

2

How This Book Works

Welcome to ChatGPT for Restaurants

This guide is designed to help you save time, streamline your restaurant operations, and enhance your customer service using ChatGPT, an advanced AI language model.

In the following chapters, you'll discover how ChatGPT can support various aspects of your restaurant management duties, from marketing and promotions to team management and customer service. Rather than just reading about these applications, we'll provide you with actionable, tested ChatGPT prompts that you can implement right away.

Each section of the book will focus on a specific area of restaurant management. We'll give you ready-to-use prompts to help you generate content or automate tasks, as well as the actual outputs from ChatGPT so you can see exactly how it works in practice. This will allow you to explore and experiment with ChatGPT directly, making it a more interactive experience.

As you explore these examples, it's important to note that my passion and experience are rooted in Thai food, so I thought it only fitting to use examples throughout this book related to a Thai restaurant and Thai dishes. However, whether you own or manage a Thai restaurant or any other type of restaurant, these prompts can easily be tweaked to meet your specific needs. If the examples don't directly apply to your cuisine, simply adjust the ingredients, menu items, or even the tone of voice to fit your brand. This book is designed to guide and inspire, but the customization is in your hands.

Here's where you can really make this book your own: these prompts are here to spark your creativity and provide a starting point, but they can be easily modified to suit your business's unique needs. Don't be afraid to get creative with them! For example, if you start by using a prompt exactly as it's written and feel the output is too generic, don't worry. You can always ask ChatGPT to adjust the response by adding your business name, specific ingredients, prices, dates, or any other detail that's unique to your restaurant. ChatGPT is designed to help you fine-tune the results, so if it's not exactly what you want on the first try, keep chatting with it until you get the output that feels right for you.

Have fun with the process! This is a tool for experimentation, and you're in control. If the output doesn't quite hit the mark, you can easily refine it. For instance, you can adjust the tone by saying:

- "Make it sound more casual and friendly."
- "Could you make the tone more formal and professional?"

- "Can you add some humor to this response without being too casual?"

You can always start over if you're not happy with the results. ChatGPT gives you the flexibility to experiment until you find exactly what you're looking for.

Alongside each example prompt, we'll offer suggestions on how to tweak and personalize the prompts for your restaurant's unique needs. These suggestions will guide you in tailoring ChatGPT's responses to better fit your brand, cuisine, and customer base.

Bonus Chapter: Advanced Prompting Techniques

Throughout this book, you'll come across certain prompts marked with the tag *"See Bonus Chapter for an Expanded Version"*. These prompts are the ones we'll explore in greater depth in the **Bonus Chapter**, where we'll provide longer, more detailed variations and advanced strategies to refine your results with ChatGPT.

In the Bonus Chapter, you'll find:

- Expanded, in-depth versions of prompts
- Advanced customizations for each prompt
- Tips on how to make your results even more tailored to your restaurant's needs

When you're ready to go beyond the basics, refer to the Bonus Chapter to take your use of ChatGPT to the next level.

One important thing to note

While we'll provide you with the output generated by ChatGPT for each prompt, it's essential to remember that you might get slightly different results when using the same prompt. This is because ChatGPT continuously evolves, and its responses can vary based on the way prompts are phrased. We encourage you to experiment with different wording and variations to get the most relevant results for your restaurant.

So, dive in, start experimenting, and remember: there are no wrong answers. The more you play with these prompts, the more you'll discover how to make them truly work for your business!

3

ChatGPT Basics for Restaurant Managers

Introduction to ChatGPT

ChatGPT is an advanced artificial intelligence language model developed by OpenAI. It's designed to understand and generate human-like text, making it a versatile tool for many tasks. Whether you're drafting an email, creating a social media post, or responding to customer feedback, ChatGPT can help. It works by processing your prompts and offering intelligent, context-aware responses that can be applied to a wide variety of restaurant-related tasks.

For restaurant and managers, ChatGPT is like having a virtual assistant that can handle many of the routine tasks that take up your valuable time. Whether it's crafting promotional content or answering customer inquiries, ChatGPT allows you to focus on higher-level responsibilities, knowing that it can handle the details with efficiency and creativity.

The Benefits for Restaurant Managers

ChatGPT can be an invaluable tool in various areas of restaurant management, including:

- **Daily Operations**: Automating routine tasks like scheduling, inventory management, and staff communication.
- **Marketing**: Generating compelling content for social media, email newsletters, and in-house promotions.
- **Customer Service**: Crafting consistent and friendly responses for customer reviews, inquiries, and feedback.
- **Team Management**: Developing training materials, sending internal updates, and assisting with team communications.
- **Financial Management**: Assisting with budgeting, expense tracking, and generating financial reports. ChatGPT can help break down complex financial data and provide insights, making it easier to manage costs, forecast revenue, and analyze profit margins.

By integrating ChatGPT into your daily workflow, you can increase efficiency, save time, and focus on providing excellent service to your guests while effectively managing your team.

Getting Started with ChatGPT

To begin using ChatGPT, here are a few essential steps:

- **Sign Up**: First, visit OpenAI's website at https://www.openai.com

and create an account to access ChatGPT. The free version is more than enough for most restaurants to get started and explore ChatGPT's capabilities. However, if you find you need more advanced features later on, you can always consider upgrading to a paid plan.

- **Access ChatGPT**: Once you've signed up, you can access ChatGPT through its web interface. This is where you'll input your prompts and receive responses. You can use it on desktops, laptops, tablets, and mobile devices for convenience.
- **Creating Prompts:** To get started, simply type your question or request into the ChatGPT input field. For example, you might ask, "Write a promotional Instagram post for a new menu item." ChatGPT will then process your request and generate a response based on its vast training data.

Tips for Crafting Effective Prompts

To get the best results from ChatGPT, here are some tips for crafting clear and specific prompts:

- **Be Direct:** The more specific you are, the more relevant the response will be. For example, instead of asking for "a social media post," try a prompt like, "Write a catchy Instagram post for our new vegan curry, using fresh ingredients and highlighting its mild flavor."
- **Provide Context:** ChatGPT works best when it knows the context of your restaurant. Include details about your menu, location, or specific event, so it can tailor responses to your needs.

- **Refine and Experiment:** If the first response isn't exactly what you wanted, refine your prompt. ChatGPT is flexible and can adjust based on the feedback you provide. Don't be afraid to experiment with different wording to fine-tune the results.

Now the Fun Begins!

Now that you understand the basics of ChatGPT and how to get started, you're ready to dive into the practical applications. In the next chapters, we'll explore how you can use ChatGPT for marketing, customer service, team management, and more. You'll also find tested and effective prompts that have been specifically designed for restaurant managers like you.

So, let's get started with some actionable prompts and discover how ChatGPT can make your daily operations easier and more efficient!

4

Marketing and Promotions

Marketing is the lifeblood of any successful restaurant. A strategic marketing plan not only attracts new customers but also keeps loyal patrons engaged and excited. With ChatGPT as your marketing assistant, you can generate dynamic content for social media, email campaigns, and promotional events—all tailored to your restaurant's unique style.

In this chapter, we'll explore how to use ChatGPT to streamline your marketing efforts, offering actionable prompts and outputs to inspire your campaigns. Each example will include suggestions for customizing the prompts to ensure they align with your brand and audience.

Using ChatGPT for Social Media

Social media platforms like Instagram, Facebook, and TikTok provide a direct line to your customers. ChatGPT can help you craft posts that resonate with your audience, promote specials, and build your brand.

Prompt 1: Instagram Post for a New Dish

Prompt: *"Write a catchy Instagram post for our new dessert: Coconut Sticky Rice with Mango. Include a description, hashtags, and a call to action."*

Output from ChatGPT

Sweet, creamy, and oh-so-delicious! 🥭✨ Our Coconut Sticky Rice with Mango is here to brighten your day. Perfectly ripe mangoes paired with silky coconut sticky rice make for the ultimate treat. 🥥
Come and indulge in this tropical delight today! 🌴
#DessertLovers #ThaiCuisine #FoodieGoals #CoconutStickyRice

Making Each Prompt Your Own: Customization Tips

To get the most relevant output from ChatGPT, provide more specific details about your restaurant, which will help ChatGPT craft a post that's perfectly suited to your needs. Here are some ways you can customize your prompt:

- **Restaurant Name**: Give ChatGPT your restaurant's name for branding, e.g., "Now available at [Restaurant Name]!"
- **Menu Details**: If your dish has unique twists or special ingredients, share them with ChatGPT, such as "Made with handpicked mangoes and fresh coconut cream."
- **Availability**: If it's a limited-time offer or a seasonal special, tell ChatGPT, e.g., "Available only until [date]—don't miss out!"
- **Target Audience**: Inform ChatGPT of your audience's preferences, whether casual or elegant, so it can adjust the tone accordingly, e.g., "Perfect for a sweet treat after a spicy meal!"
- **Location and Dining Options**: If applicable, provide ChatGPT with details like "Order now for delivery or pickup at [restaurant address]."
- **Promotions or Offers**: If you're running a special offer like a discount or bundle, mention it to ChatGPT so it can integrate that into the post, e.g., "Get 10% off when you order this dessert with any entrée!"

By providing ChatGPT with these details, you'll ensure the generated Instagram post aligns with your restaurant's specific needs, style, and customer base.

Prompt 2: TikTok Script for a Behind-the-Scenes Video

"See Bonus Chapter for an Expanded Version"

Prompt: *"Create a TikTok script showcasing how we prepare our signature Pad Thai. Include engaging captions and trending hashtags."*

Output from ChatGPT

🎥 Opening shot: Chef tossing noodles in a sizzling wok.
🎤 Voiceover: "Ever wondered how our famous Pad Thai comes to life? Let's take you behind the scenes!"
🍴 Caption: "A symphony of flavors, cooked to perfection. Ready to try? Come visit us! 🌟"
🎶 Hashtags: #BehindTheScenes #ThaiFoodLove #PadThaiPerfection

Making Each Prompt Your Own: Customization Tips

To make this TikTok script truly aligned with your restaurant's brand and target audience, here's how you can customize the prompt for ChatGPT:

- **Dish Details**: Provide ChatGPT with specifics about your signature dish, such as "Our Pad Thai features freshly made rice noodles, local vegetables, and a secret blend of spices." This helps the script reflect what makes your dish unique.

- **Chef or Staff Appearance**: Mention who's in the video—whether it's a head chef, sous chef, or another team member—and any notable personalities or faces that customers might recognize. For example, "Chef [Name] shows us how it's done!"
- **Engagement Tactics**: If you want to encourage interaction, ask ChatGPT to include a specific call-to-action (CTA), like "Tag a friend who needs to try this Pad Thai!" or "Mention this video for 10% off your meal today!"
- **Bonus Chapter:** Be sure to explore the Bonus Chapter for a deeper guide on how to customize this prompt, ensuring you get more detailed and tailored responses from ChatGPT.

By feeding these specifics to ChatGPT, you ensure that the TikTok video script fits your restaurant's unique qualities, helping it stand out and resonate more with your audience.

Prompt 3: Facebook Post for a Limited-Time Offer

Prompt: *"Generate a Facebook post for a 'Kids Eat Free' promotion every Sunday."*

Output from ChatGPT

👨‍👩‍👧‍👦 Family Sundays just got sweeter! Every Sunday, kids eat free with the purchase of an adult main. ✏️ Make it a day to remember with delicious food and family fun!

Reserve your table today at [website link]. 📅 See you Sunday! 🎉"

Making Each Prompt Your Own: Customization Tips

To ensure the 'Kids Eat Free' promotion is tailored specifically to your restaurant's needs, provide ChatGPT with these additional details:

- **Age Limits or Terms**: If you have specific terms (like age limits), share them with ChatGPT. For example, "Kids under 12 qualify" or "Free meal with the purchase of an adult entrée per child."
- **Highlight Kids' Menu Items**: Mention the popular kids' menu items that should be featured in the post, such as "Kids can choose from favorites like mac 'n cheese, mini burgers, or our signature chicken tenders."
- **Additional Details**: If the promotion is tied to any other details, such as time limits or exclusions, be sure to include those in your prompt. For example, "Offer valid from 12 PM to 5 PM only."
- **Visual Content**: Let ChatGPT know if you'd like to include specific types of imagery, like "Add a family-friendly photo of kids enjoying meals at our restaurant" or "Feature our chef preparing kids' dishes to emphasize freshness."
- **Location Information**: If applicable, add location-based details like "Join us at [Restaurant Name] at [location], your neighborhood favorite for family dining!"
- **Tone Adjustment**: If your restaurant has a specific atmosphere (e.g., casual, upscale, or themed), guide ChatGPT on adjusting the tone. For instance, if your restaurant is family-friendly and casual, the tone should be warm and inviting, while a more formal setting might prefer a sophisticated message.

By providing these specifics, ChatGPT can refine the post to align perfectly with your promotion and brand, making it more relevant and engaging for your audience.

Prompt 4: Instagram Carousel for Seasonal Specials

Prompt: *"Create captions for a 5-image Instagram carousel featuring our winter menu specials."*

Output from ChatGPT

Slide 1: "❄ Winter is here, and so is our new menu! 🌟 Let's explore what's cooking…"

Slide 2: "🍲 Cozy up with our Spicy Tom Yum Soup—perfect for chilly nights!"

Slide 3: "🥢 Warm your soul with our hearty Thai Green Curry."

Slide 4: "🍨 End on a sweet note with our Coconut Sticky Rice with Mango."

Slide 5: "🌟 Which one will you try first? Tag your foodie friends and let us know! 👇"

Making Each Prompt Your Own: Customization Tips

To make this Instagram carousel even more aligned with your restaurant's identity and offerings, you can customize the prompt by providing ChatGPT with the following details:

- **Specific Dishes**: Share the names of the dishes you want to highlight, especially if your winter specials differ from the ones mentioned in the example. For instance, "Replace the Tom Yum Soup with our signature Spicy Ramen, perfect for a winter evening!"
- **Ingredient Sourcing**: If you pride yourself on using locally sourced or organic ingredients, provide ChatGPT with that information. For example, "Highlight our locally sourced organic vegetables in the caption for our Thai Green Curry."
- **Seasonal Items**: Let ChatGPT know if you're promoting seasonal or limited-time dishes so that the urgency comes through in the copy. You can say, "Make sure to emphasize that our winter specials are available only through February."
- **Prices for Transparency**: If you want to provide your audience with pricing details, add that to your prompt, like "Include price ranges for each dish: $12–$20."
- **Restaurant's Unique Voice**: Adjust the tone of the captions based on your restaurant's vibe. If your restaurant has a cozy, family-friendly atmosphere, you may want a warm, inviting tone. If your place is more trendy, you might opt for something more playful or cutting-edge. For example, "Slide 2: 'Cozy up and spice up your winter with our heartwarming Spicy Ramen—only available this season!'"
- **Add a Call-to-Action (CTA)**: Encourage engagement by specifying

a CTA you want. For example, "Invite followers to vote on their favorite winter dish or share a winter memory to connect with your community." Or, "Encourage them to reserve a table through the link in bio: 'Don't miss out—book your spot today!'"

- **Visual Customization**: If you'd like ChatGPT to mention any specifics about images, provide details like "Include images of steaming dishes and a cozy ambiance in the background" or "Showcase a close-up of the ingredients to highlight freshness."

By offering ChatGPT this extra information, you can get even more customized and impactful captions that reflect your restaurant's seasonal specials and unique brand.

Email Marketing and Newsletters

Email campaigns help keep your audience informed and engaged. Use ChatGPT to write persuasive and value-packed messages for your subscribers.

Prompt 5: Event Invitation Email

Prompt: *"Write an email inviting customers to a wine pairing night featuring Thai cuisine. Include event details and RSVP instructions."*

Output from ChatGPT

Subject Line: "Wine Meets Thai: An Evening of Flavor Awaits!"

"Dear [Customer's Name],

Join us on [date] for a special Wine Pairing Night at [Restaurant Name]. Experience the magic of Thai cuisine paired perfectly with hand-selected wines. 🍷🥢

✨ Event Details:

Date: [date]

Time: [time]

Location: [restaurant address]

Spaces are limited—reserve your spot now by calling [phone number] or visiting [website]. Don't miss this one-of-a-kind culinary event!

Cheers,

[Your Name]"

Making Each Prompt Your Own: Customization Tips

To ensure the event invitation email truly reflects your restaurant's unique offerings, provide ChatGPT with more specific information by considering the following customizations:

- **Event Details**: Give ChatGPT precise details about the wine pairing night, such as the wines featured. For example, "Mention our exclusive selection of organic wines, including a local red from [vineyard name]."
- **Descriptions of Dishes and Wines**: Add some background on the dishes or the wines to make the invitation more appealing. For

instance, "Highlight the fact that the Pad Thai will be paired with a refreshing white wine to bring out the perfect balance of flavors."

- **Discounts or Special Offers**: If you're offering early-bird discounts or exclusive VIP packages, let ChatGPT know. For example, "Include an early-bird special—10% off for those who reserve by [date]."
- **Special Guests**: If you have a sommelier, chef, or any special guest participating, include this information to create exclusivity. You can say, "Mention that renowned sommelier [Name] will guide guests through the pairings."
- **RSVP Instructions**: Customize the RSVP instructions based on your restaurant's reservation system or preferences. You can say, "Add a line directing people to RSVP via our website, or if we're using a reservation system like Resy, mention that option."
- **Tone of the Invitation**: Ensure the tone matches your brand—whether it's upscale, casual, or cozy. If your restaurant has a formal atmosphere, the email should reflect that with a polished, refined tone. Alternatively, if your place is more laid-back, make the tone friendly and approachable. You can specify: "Adjust the tone to match our fine-dining style—formal but inviting."
- **Call-to-Action (CTA)**: Provide ChatGPT with a specific CTA you want to include in the email. For example, "Encourage recipients to RSVP as soon as possible, mentioning the limited seating, or to share the event with friends."

By providing ChatGPT with these specifics, you'll receive a much more tailored email that reflects the unique aspects of your restaurant's wine pairing event and captures the attention of your customers.

Prompt 6: Loyalty Program Email

Prompt: *"Write an email introducing a new loyalty program for regular customers."*

Output from ChatGPT

Subject Line: "Introducing Our Loyalty Program: Rewards You'll Love!"

"Dear [Customer's Name],

We're thrilled to introduce our new Loyalty Program—your ticket to exclusive rewards and special perks! 🎉

Here's how it works:

Earn points for every dollar spent.

Redeem points for discounts, free dishes, and more!

Sign up today and receive 50 bonus points on your first visit. ✏️ Let's make every meal more rewarding!

[Sign Up Here]"

Making Each Prompt Your Own: Customization Tips

To get the best results and make the loyalty program email truly fit your restaurant's needs, provide ChatGPT with the following details:

- **Personalization**: Add personal touches to your loyalty program email by specifying the customer's name and even their past favorites. For example: "Use their first name in the greeting and mention green curry chicken, a popular dish they've ordered before."

- **Program Structure**: Describe the points system in more detail, such as how many points are earned per dollar and what rewards are offered. For example, "Mention that customers earn 1 point per dollar spent, with 100 points earning a free appetizer."
- **Milestone Rewards**: If your program has tiered rewards or milestones, let ChatGPT know what they are. You could say, "Include information about milestone rewards like 'free dessert after 100 points' or '20% off after 500 points.'"
- **Visual Elements**: If you have visuals, such as mockups of your loyalty card or digital rewards system, mention them so ChatGPT can suggest ways to highlight these in the email. For example: "Incorporate an image of our loyalty card design in the email, showing customers how to track their rewards."
- **Program Name**: Give your loyalty program a creative, branded name that reflects your restaurant's personality. For example: "Suggest a name for the program like 'The Spice Club' or 'Golden Mango Rewards,' which ties in with our Thai theme."
- **Exclusive Offers**: If you offer special sign-up bonuses, like extra points or exclusive discounts for new members, be sure to let ChatGPT know. For example: "Add a bonus for new sign-ups—'Get 50 points just for joining today!'"
- **Tone and Call-to-Action**: Specify the tone of the email (friendly, formal, etc.) based on your brand's voice. You can also provide clear CTAs like, "Emphasize 'Sign up today to start earning!' or 'Join our loyalty program now to enjoy perks right away!'"

By giving ChatGPT these specifics, you can craft a loyalty program email that's not only engaging but also directly tailored to your restaurant's unique offerings and audience.

Seasonal and Event Promotions

Special events and seasonal campaigns keep your marketing fresh and exciting. ChatGPT can help craft posts that capture your customers' attention.

Prompt 7: Holiday Catering Promotion

Prompt: *"Create a promotional post for holiday catering services."*

Output from ChatGPT

🎄 Hosting a holiday party? Let us take care of the food! 🎅 Our catering menu features all your Thai favorites, perfect for gatherings big or small. 🥢

Book your holiday catering now and enjoy stress-free celebrations. Call us at [phone number] or visit [website]. ✨

Making Each Prompt Your Own: Customization Tips

To ensure the catering post perfectly fits your restaurant's offerings, provide ChatGPT with the following specifics:

- **Menu Details**: If you offer special catering dishes or packages, provide ChatGPT with a list to feature. For example, "Mention our holiday catering specials like 'Curry Feast for 10' or 'Thai Appetizer Sampler.'"
- **Special Offers**: If you plan to offer any holiday discounts or promotions, let ChatGPT know. For instance, "Add a 15% discount for orders placed before [date]," or "Offer free delivery for orders over $200."
- **Additional Services**: If you offer added services like setup or personalized menus, be sure to mention these details. For example, "Include a line about 'Customized menus for your holiday event' or 'Free setup for large catering orders.'"
- **Delivery & Pickup**: Clarify any delivery options, especially if you're offering local delivery or pick-up services. For example, "Include delivery details like 'We deliver within [zip code range]' or 'Order online for convenient pickup at [restaurant address].'"
- **Contact Information**: Ensure ChatGPT is given your business name, website, and phone number for easy reference in the post. For example, "Don't forget to add our contact info: 'Call [phone number] or visit [website] to book your holiday feast!'"
- **Urgency**: If applicable, mention any deadlines or booking requirements. For example, "Let customers know, 'Order by [date] for guaranteed delivery on [holiday].'"

By providing these details, you ensure that ChatGPT can generate a catering post that's clear, compelling, and tailored to your business.

Prompt 8: Anniversary Celebration Post

"See Bonus Chapter for an Expanded Version"

Prompt: *"Write a post celebrating the restaurant's 5th anniversary and offering a special discount."*

Output from ChatGPT

🎉 We're Turning 5! 🎉

Thank you for five amazing years of delicious food and wonderful memories. To celebrate, we're offering 20% off your meal this week! 🥢 Join us from [start date] to [end date] and let's celebrate together.

Here's to many more years of great food and great company! ❤️

Making Each Prompt Your Own: Customization Tips

To ensure the anniversary post reflects your restaurant's unique story, provide ChatGPT with these personalized details:

- **Discount Details**: Specify the exact discount or offer you'd like to promote. For example, "Offer a 'Buy one, get one free dessert' deal or '15% off any entrée.'"

- **Event-Specific Information**: If you're hosting any special events for the anniversary, provide that info for ChatGPT to incorporate. For example, "Mention our anniversary party on [date], featuring live music or a special guest chef."
- **Visual Content**: If you have photos, videos, or a visual timeline of your restaurant's journey, provide ChatGPT with this for inclusion. For example, "Ask for a mention of our highlight reel: 'Check out our 5-year journey on Instagram!'"
- **Personalized Messaging**: Share your restaurant's history or any memorable milestones to make the message feel authentic. For instance, "Ask ChatGPT to mention how we started as a family-run business and now we've served over [number] guests!"
- **Hashtags & Branding**: If you're creating a unique hashtag for the anniversary or a specific promotion, share it with ChatGPT. For example: "Include the hashtag #FiveYearsOfFlavor or #CelebratingWith[RestaurantName]." Or ask ChatGPT to come up with a list of potential hashtags for you.
- **Sense of Urgency**: If your anniversary offer is time-sensitive, let ChatGPT know. For example, "Specify 'Offer valid from [start date] to [end date] only!' to create a sense of urgency."
- **Customer Appreciation**: Include a thank-you message for your loyal customers. For example, "Let customers know, 'We couldn't have done it without you!'"
- **Bonus Chapter**: Remember the Bonus Chapter offers more insights on how to experiment with and expand this prompt.

By providing these specifics, you'll be able to craft a more engaging, authentic, and personalized anniversary post that aligns with your restaurant's story and brand.

Prompt 9: Contest Announcement Post

Prompt: *"Write a post announcing a photo contest for customers to win a free meal."*

Output from ChatGPT

"📸 Contest Alert! Show us your love for Thai food and win a FREE meal! 🥢

Here's how to enter:

1️⃣Snap a photo of your favorite dish at [Restaurant Name].

2️⃣Post it on Instagram with #TasteOf[RestaurantName].

3️⃣Tag us and your foodie friends!

The best photo wins a free meal for two. Contest ends [date]. Good luck!"

Making Each Prompt Your Own: Customization Tips

To ensure this contest post is tailored to your restaurant and promotion, provide ChatGPT with these specifics:

- **Prizes & Runner-Up Rewards**: Mention any additional prizes or incentives. For example, "Offer a $25 gift card for the second-place winner" or "Include a free appetizer for all runners-up."
- **Signature Dishes or New Items**: If you want to highlight specific menu items, let ChatGPT know. For instance, "Encourage

customers to take photos of our Pad Thai or new Mango Sticky Rice dessert for a chance to win."

- **Unique Hashtags**: Create a memorable and catchy hashtag for your contest, and ensure it's easy for participants to remember. For example, "Ask ChatGPT to include something like #SpiceUpYourLifeAt[RestaurantName] or #MyThaiMoment."
- **Contest Dates**: Specify the contest start and end dates so customers know when to enter. For example, "Mention the contest will run from [start date] to [end date]."
- **Social Media Handles**: Provide your exact Instagram handle, as well as any other relevant social media accounts, so ChatGPT can incorporate these into the post. For example, "Make sure to tag us @YourRestaurantName on Instagram!"
- **Additional Instructions**: Clarify any specific guidelines or requests for participants, such as "Make sure the photo includes our signature Thai iced tea" or "Show us your best shot of our outdoor seating area."
- **Location-Based Instructions**: If your contest is specific to a particular location, be sure to clarify that. For instance, "If the contest is only for customers at a certain location, let ChatGPT know: 'Enter the contest at our [Location] branch only!'"

By providing these details, ChatGPT can help craft a post that speaks to your unique restaurant experience, engages your audience, and drives more participation.

Prompt 10: Community Event Collaboration Post

Prompt: *"Write a post promoting a partnership with a local charity for a fundraising dinner."*

Output from ChatGPT

🌟 Dine for a Cause! 🌟

We're partnering with [Charity Name] to host a special fundraising dinner on [date]. A portion of the proceeds will go toward [cause]. Join us for an evening of great food and giving back to our community.

Reserve your table today at [website link] or call [phone number]. Together, we can make a difference! 💖

Making Each Prompt Your Own: Customization Tips

To ensure this community-focused post aligns with your restaurant and charity event, share the following details with ChatGPT:

- **Charity's Mission**: Provide background on the charity you're supporting. For example, "Share the charity's mission, like 'A portion of proceeds will go toward providing meals for underprivileged children.'"
- **Event Details**: Give ChatGPT the exact time, location, and date of the event. For instance, "Specify the fundraising dinner will be held at [Restaurant Name] on [Date] from [Time] to [Time]."

- **Target Goal**: If your event has a fundraising goal, let ChatGPT know so it can be mentioned in the post. For example, "We hope to raise $500 for [Cause]."
- **Past Event Highlights**: If you've partnered with charities in the past, share any photos or highlights to include. For instance, "Mention previous successful events or showcase a photo from last year's charity dinner."
- **Specific Call-to-Action**: Encourage people to reserve a spot by providing your website link, phone number, and any special instructions. For example, "Make sure to include the line 'Reserve your table now at [website] or call [phone number] to support this great cause!'"
- **Add Incentives**: If you plan to offer something extra, such as a discount or complimentary drink for attendees, make sure ChatGPT includes that. For example, "Add 'Guests will receive a complimentary glass of wine when they join us for the event.'"
- **Visuals**: If you have any event photos or the charity's logo, share them with ChatGPT for inclusion. For example, "Ask for a mention of our event flyer or the charity's logo in the post."

By providing these specifics, ChatGPT can help generate a well-rounded, impactful post that promotes both the community cause and your restaurant's role in making a difference.

Conclusion

With these 10 marketing prompts and your creative input, ChatGPT can become your personal assistant in crafting compelling campaigns, engaging social media posts, and effective promotions. Remember, the outputs provided here are just starting points—by experimenting and tailoring them, you can create content that resonates deeply with your audience and reflects your restaurant's unique brand.

Now, let's move to Chapter 5, where we'll explore how ChatGPT can enhance **customer engagement** and streamline communication.

5

Customer Engagement

Customer engagement is at the heart of building loyalty and fostering meaningful relationships with your patrons. By leveraging ChatGPT, you can respond to inquiries, address concerns, and celebrate special moments with customers, all while maintaining a personal touch. This chapter provides actionable prompts to help you use ChatGPT to streamline customer interactions and create memorable experiences.

We'll include 10 prompts with sample outputs and tips for tailoring each one to your specific restaurant needs. These examples will showcase how ChatGPT can enhance your communication strategies while saving you time.

Automating Customer Responses

Efficient and empathetic communication is essential for handling customer inquiries and feedback. ChatGPT can craft professional, friendly, and consistent responses that align with your brand voice.

Prompt 1: Apologizing for Long Wait Times

Prompt: *"Write a response to a customer complaint about wait times, offering an apology and a discount for their next visit."*

Output from ChatGPT

"Dear [Customer's Name],

Thank you for bringing this to our attention. We sincerely apologize for the extended wait time during your recent visit. This is not the level of service we strive to provide, and we're committed to improving.

As a token of our apology, we'd like to offer you a 15% discount on your next visit. Please mention this message when you arrive, and we'll ensure a better experience.

We appreciate your understanding and hope to see you again soon.

Best regards,

[Your Name]"

Making Each Prompt Your Own: Customization Tips

To personalize this response and make it more relevant to your restaurant, give ChatGPT these additional details:

- **Incident Details**: Share specifics about the situation. For example, "If the complaint was about a long wait on a busy Saturday night, mention that: 'We're sorry for the delays during your visit on Saturday evening—it was an unusually busy time.'"
- **Custom Discount Offers**: Adjust the discount to align with your restaurant's policy. For instance, "Provide ChatGPT with the percentage discount you offer, or suggest alternatives like a free appetizer or dessert."
- **Corrective Measures**: Reassure the customer by outlining steps you're taking to prevent the issue in the future. For example, "Mention, 'We're adding more staff during peak hours to ensure smoother service.'"
- **Tone Adjustment**: If your restaurant's voice is more casual or formal, adjust the tone accordingly. For instance, "Change the wording to fit your style, such as 'We're so sorry you had to wait—it's not cool, and we'll do better!'"
- **Customer Interaction Preferences**: Let ChatGPT know if you prefer offering a direct follow-up, such as "Feel free to reach out to me directly at [email/phone number] if you have more concerns."

Providing these details ensures the apology feels sincere, specific, and uniquely reflective of your restaurant's commitment to great service.

Prompt 2: Responding to a Positive Yelp Review

Prompt: *"Generate a response to a positive review left on Yelp."*

Output from ChatGPT

"Hi [Reviewer's Name],

Thank you so much for your kind words! We're thrilled to hear you enjoyed [specific dish or experience mentioned in the review]. Our team works hard to create memorable dining experiences, and your feedback means the world to us.

We can't wait to welcome you back for another visit. Let us know next time you're in—we'd love to thank you in person!

Warm regards,

[Your Name]"

Making Each Prompt Your Own: Customization Tips

To make this response more tailored to your restaurant and customer feedback, provide ChatGPT with:

- **Details from the Review**: Highlight specific dishes or services the customer praised. For example, "If they mentioned loving your Tom Yum Soup, say, 'We're so glad you enjoyed our Tom Yum Soup—it's one of our favorites too!'"
- **Upcoming Features or Events**: Use the opportunity to invite the

reviewer back. For instance, "Add, 'We're launching a new seasonal menu soon—we think you'll love the Duck and Lychee Curry!'"

- **Loyalty Incentives**: Encourage repeat visits by mentioning perks. For example, "Invite them to join your loyalty program: 'Next time, be sure to sign up for our Mango Rewards program for exclusive perks!'"
- **Tone of Response**: Match your brand's voice. For example, "Adjust the tone to align with your style, whether it's casual ('You rock! Thanks for the awesome review!') or formal ('We deeply appreciate your kind feedback.')."
- **Special Recognition**: Let ChatGPT know if you want to add an extra touch, like a shout-out to the team. For example, "Add, 'Our server Sarah will be thrilled to hear you enjoyed her service!'"

Sharing these specifics helps ChatGPT craft a heartfelt and personal response that deepens the connection with your customers while promoting future visits.

Creating Personalized Messages

Personalization can transform routine interactions into moments of connection. ChatGPT can help you craft messages for birthdays, anniversaries, and other special occasions.

Prompt 3: Birthday Greetings for a Loyal Customer

Prompt: *"Write a personalized birthday message for a frequent customer."*

Output from ChatGPT

"Happy Birthday, [Customer's Name]! 🎉

Thank you for being a valued member of our [Restaurant Name] family. To celebrate your special day, we'd like to treat you to a complimentary dessert during your next visit. Just let us know when you're stopping by—we can't wait to celebrate with you!

Wishing you a fantastic year ahead.

Warm regards,

[Your Name]"

Making Each Prompt Your Own: Customization Tips

To create a birthday message that feels truly personal, share additional details with ChatGPT:

- **Favorite Items**: Mention their favorite dish, drink, or dessert to make the message extra special. For example, "Happy Birthday, [Customer's Name]! We'd love to treat you to your favorite dessert, our famous Mango Sticky Rice, on your next visit!"
- **Additional Perks**: Consider offering more than a dessert, such as a birthday discount or a special appetizer. Let ChatGPT know,

for instance, "Add: 'Enjoy 15% off your meal when you visit us this week!'"

- **Custom Tone**: Adjust the tone to reflect your restaurant's branding. For example, "For a playful tone, say, 'Hope your birthday is as sweet as our desserts!'"
- **Call to Action**: Include practical details, such as how they can redeem the offer. For instance, "Add: 'Mention this message when making your reservation to claim your treat!'"
- **Personal Connections**: If your team has a close relationship with the customer, personalize even further: "Add: 'We all wish you the happiest birthday—[Chef's Name] and the team can't wait to see you!'"

Providing specific information ensures the birthday message aligns perfectly with your customer's preferences and your restaurant's style.

Prompt 4: Anniversary Congratulations

"See Bonus Chapter for an Expanded Version"

Prompt: *"Write a congratulatory message for a couple celebrating their anniversary at the restaurant."*

Output from ChatGPT

"Dear [Customer's Name],

Happy Anniversary to you and [Partner's Name]! 💕 We're honored that you chose to celebrate this special milestone with us. To make your evening even more memorable, we'd like to offer a complimentary glass of wine or dessert for the two of you.

Cheers to many more years of love and happiness!

Warm wishes,

[Your Name]"

Making Each Prompt Your Own: Customization Tips

To craft an anniversary message that feels unique and thoughtful, share the following details with ChatGPT:

- **Pre-Arranged Surprises**: If the couple has planned something specific, incorporate it into the message. For example, "Add: 'We've prepared the romantic table you requested, complete with candlelight and flowers.'"
- **Custom Offers**: Tailor the complimentary offer to fit your policies. For example, "Change the offer to 'a complimentary champagne toast' or 'a shared dessert platter.'"
- **Restaurant Branding**: Adjust the tone to suit your restaurant's ambiance. For instance, "For an upscale tone, say, 'We are delighted to share in your celebration at [Restaurant Name], where love and fine dining come together.'"
- **Highlight Romantic Features**: Emphasize your restaurant's

romantic elements, such as scenic views, live music, or cozy corners. For example, "Add: 'Enjoy a romantic evening with a view of the city skyline and live violin music.'"

- **Encourage Sharing**: Invite the couple to share their special moment. For example, "Add: 'Tag us in your anniversary photos—we'd love to celebrate with you!'"
- **Bonus Chapter**: Check out the Bonus Chapter for a detailed guide on expanding this prompt and customizing it to fit your restaurant's unique needs.

Providing these details helps ChatGPT craft a congratulatory message that feels personal, aligned with your restaurant's theme, and designed to make the couple's celebration even more special.

Handling Negative Reviews

Addressing negative reviews with professionalism can turn a dissatisfied customer into a loyal one. ChatGPT can help you craft thoughtful and empathetic responses.

Prompt 5: Responding to a Negative Review About Food Quality

Prompt: *"Create a thoughtful response to a negative review about food quality."*

Output from ChatGPT

"Dear [Reviewer's Name],

We're truly sorry to hear that your recent experience didn't meet your expectations. At [Restaurant Name], we pride ourselves on delivering high-quality dishes, and it's disappointing to know we fell short.

We value your feedback and would like to make it right. Please reach out to us directly at [contact information] so we can address your concerns and ensure a better experience next time.

Thank you for taking the time to share your thoughts—we hope to have the opportunity to serve you again soon.

Best regards,

[Your Name]"

Making Each Prompt Your Own: Customization Tips

To ensure your response feels personalized and genuine, provide ChatGPT with additional details:

- **Acknowledge Specific Complaints**: If the reviewer mentioned a particular dish or issue, include it in your response. For example, "Add: 'We're sorry the Pad Thai didn't meet our usual standards.'"

- **Offer Solutions**: Let ChatGPT include specific actions you're taking to resolve the issue, such as a refund, replacement meal, or a complimentary visit. For instance, "Add: 'We'd love to invite you back to try a fresh serving, on the house.'"
- **Tone Customization**: Specify the tone you'd like ChatGPT to use, such as empathetic, professional, or casual, depending on your brand. For example, "Make the tone more heartfelt, like: 'We're deeply sorry for the disappointment and truly value your patronage.'"
- **Proactive Follow-Up**: Include an invitation to discuss the matter further. For instance, "Add: 'We'd appreciate a chance to hear more about your experience so we can improve.'"
- **Highlight Improvements**: If you've made changes based on feedback, mention them. For example, "Add: 'We've since adjusted our recipe to ensure it meets our usual quality.'"

By incorporating specific feedback and solutions, ChatGPT can craft a response that not only addresses the issue but also leaves the reviewer feeling heard and valued.

Prompt 6: Following Up After Resolving a Complaint

Prompt: *"Write a follow-up message to a customer after resolving their complaint."*

Output from ChatGPT

"Hi [Customer's Name],

Thank you for giving us the opportunity to make things right. We're so glad we could address your concerns and appreciate your understanding.

Your feedback helps us grow, and we're committed to providing you with the exceptional dining experience you deserve. Please don't hesitate to reach out if there's anything else we can do for you.

Looking forward to welcoming you back soon!

Warm regards,

[Your Name]"

Making Each Prompt Your Own: Customization Tips

To make follow-up messages feel personalized and relevant, provide ChatGPT with these extra details:

- **Mention the Resolution**: Include the specific way you addressed their complaint. For example, "Add: 'We're glad you enjoyed the replacement dish we prepared for you.'"
- **Encourage a Return Visit**: ChatGPT can include an invitation, such as a special offer. For example, "Add: 'We'd love to welcome you back soon and treat you to a complimentary appetizer.'"
- **Personalize Based on Context**: If you know the customer's preferences, use them. For instance, "Add: 'Next time, let us know if you'd like the Green Curry extra spicy—just the way you like it!'"
- **Set a Warm Tone**: Specify a tone that matches your restaurant's

brand. For example, "Make it more inviting: 'We're excited to see you again soon for another delicious meal!'"

- **Highlight Changes Made**: Mention any improvements resulting from their feedback. For example, "Add: 'Your comments inspired us to retrain our staff to ensure consistent quality.'"

Providing these details allows ChatGPT to create a follow-up message that reinforces your dedication to exceptional customer service and builds trust with your guests.

Encouraging Customer Feedback

Customer feedback is invaluable for improving your services and offerings. ChatGPT can help you encourage more customers to share their thoughts.

Prompt 7: Requesting Feedback After a Meal

Prompt: *"Write a message asking a customer for feedback after dining at the restaurant."*

Output from ChatGPT

"Dear [Customer's Name],

Thank you for dining with us at [Restaurant Name]! We hope you enjoyed your experience. Your feedback means a lot to us and helps us improve.

Would you mind sharing your thoughts in a quick survey? It'll only take a minute: [insert link].

As a thank-you, you'll receive a 10% discount on your next visit!

We appreciate your time and hope to see you again soon.

Best regards,

[Your Name]"

Making Each Prompt Your Own: Customization Tips

To tailor feedback requests to your restaurant's style and offerings, consider these adjustments:

- **Include Specific Survey Tools**: Provide ChatGPT with a direct survey link or mention if you'd like a QR code integrated. For example, "Add: 'Scan the QR code on your receipt or click here: [survey link].'"
- **Adjust the Incentive**: Modify the reward based on your budget or goals. For instance, "Change to: 'As a thank-you, we'll treat you to a free appetizer on your next visit!'"
- **Personalize the Greeting**: Provide ChatGPT with customer details if available, such as their preferred dish or table. For example, "Add: 'We hope you enjoyed your Pad Thai!'"

- **Emphasize Convenience**: Highlight how quick and easy the process is. For instance, "Add: 'It's just 3 questions—your time is important to us!'"
- **Follow-Up Option**: Mention a way to follow up if they have specific concerns. For example, "Add: 'If there's anything we can improve, feel free to email us directly at [email address].'"
- **Reinforce Your Brand Voice**: Ensure the tone reflects your restaurant's personality. For example, "Make it more playful: 'Got a minute? Share your thoughts and snag a sweet deal for next time!'"

Providing these additional details ensures that ChatGPT generates a feedback request that feels genuine, engaging, and perfectly suited to your restaurant's branding and customer relationships.

Promoting Customer Retention

Engaging with customers post-visit can strengthen loyalty. Use ChatGPT to create follow-up messages that encourage repeat business.

Prompt 8: Thank-You Email for First-Time Visitors

Prompt: *"Write a thank-you email to a first-time visitor, inviting them back for another meal."*

Output from ChatGPT

"Dear [Customer's Name],

Thank you for visiting [Restaurant Name]—we loved having you! We hope you enjoyed [mention popular dish or experience].

To show our appreciation, we'd like to offer you 10% off your next visit. Just mention this email when you stop by.

Looking forward to serving you again soon!

Warm regards,

[Your Name]"

Making Each Prompt Your Own: Customization Tips

- **Personalize Based on Their Visit**: Include their specific order or a popular dish they tried. For instance, "We hope you enjoyed our signature Green Mango and Prawn Salad!"
- **Highlight Upcoming Specials**: Mention seasonal menu items or events. Example: "Don't miss our upcoming Wine Pairing Night on Friday!"
- **Invite Them to Join Your Loyalty Program**: Add, "Sign up for our rewards program and earn points toward free dishes and discounts."

- **Adjust the Incentive**: Change the offer to align with your restaurant's budget or goals. Example: "Enjoy a complimentary dessert on your next visit."
- **Reflect Your Brand's Voice**: Whether formal or playful, adapt the tone. Example: "We can't wait to welcome you back for another round of deliciousness!"

Prompt 9: Celebrating Customer Milestones

Prompt: *"Write a message congratulating a customer on reaching a loyalty program milestone."*

Output from ChatGPT

"Hi [Customer's Name],

Congratulations on reaching [milestone, e.g., 100 points] in our loyalty program! 🎉 As a token of our gratitude, we'd like to offer you [reward, e.g., a free appetizer or dessert] during your next visit.

Thank you for your continued support—we're lucky to have you as part of the [Restaurant Name] family.

Warm regards,

[Your Name]"

Making Each Prompt Your Own: Customization Tips

- **Add a Personalized Touch**: Mention their favorite dish or specific visit history. Example: "We love seeing you enjoy our Massaman Beef Curry!"
- **Show Progress Toward the Next Milestone**: "You're just 50 points away from earning a $10 reward!"
- **Make the Reward Unique**: Offer something exclusive, like early access to new menu items.
- **Include a Milestone Celebration Theme**: "Let's celebrate your achievement with a complimentary dessert during your next visit!"
- **Use Fun Visuals**: Provide ChatGPT with an idea for graphics or imagery, such as a digital badge or certificate.

Prompt 10: Inviting VIP Customers to an Exclusive Event

"See Bonus Chapter for an Expanded Version"

Prompt: *"Write a personalized invitation for a VIP customer to attend an exclusive tasting event."*

Output from ChatGPT

"Dear [Customer's Name],

You're invited to an exclusive tasting event at [Restaurant Name]! ✨ Join us on [date] for an evening of fine dining, featuring new dishes curated by our chefs.

Spaces are limited, so please RSVP by [date] to secure your spot. We'd be honored to have you join us for this special occasion.

Looking forward to seeing you there!

Warm regards,

[Your Name]"

Making Each Prompt Your Own: Customization Tips

- **Include Details About the Event**: Highlight the featured dishes, wines, or chefs. Example: "Savor a 5-course menu, including our new Red Duck Curry."
- **Offer Priority Access for VIPs**: "RSVP by Friday to guarantee your exclusive spot."
- **Highlight Exclusivity**: Emphasize that it's a private event with limited seating.
- **Include a Thank-You for Loyalty**: Add, "As one of our most valued customers, we'd love to treat you to this special experience."
- **Make the Invitation Stand Out**: Use ChatGPT to suggest phrases or visuals that reflect elegance and exclusivity, such as gold-embossed designs or formal language.
- **Bonus Chapter:** Refer to the Bonus Chapter for a deeper dive into expanding the prompt and tailoring it to your restaurant's specific needs.

Conclusion

With these 10 prompts, you can elevate your **customer engagement** strategies and foster lasting connections with your patrons. By using ChatGPT to craft thoughtful, personalized messages, you'll save time while ensuring each interaction reflects your restaurant's unique values. Experiment with these ideas to make every customer feel valued and appreciated.

6

Operational Efficiency

Running a restaurant involves juggling a multitude of tasks, from managing inventory to ensuring smooth staff communication and providing the necessary training to your team. Streamlining operations is essential to reduce errors, save time, and maintain a consistent and high-quality customer experience. With the help of ChatGPT, restaurant managers can simplify many of these tasks, from keeping track of stock levels to improving staff coordination. This chapter will provide you with actionable prompts to improve inventory management, communication, scheduling, and staff training, all aimed at boosting your **operational efficiency**.

By integrating AI-powered solutions into your operations, you can enhance workflow efficiency, boost team morale, and ensure that every aspect of your restaurant runs smoothly.

Inventory Management

Effective inventory management ensures you never run out of essential ingredients while preventing waste. ChatGPT can help you automate weekly stock checks, track expiration dates, and create reminders to reorder ingredients, saving you time and effort.

Prompt 1: Weekly Inventory Checklist for Perishables

"See Bonus Chapter for an Expanded Version"

Prompt: *"Create a weekly inventory checklist for perishables in a Thai restaurant."*

Output from ChatGPT

"Weekly Inventory Checklist – Perishables"

Fresh Vegetables:

- Thai basil
- Lemongrass
- Kaffir lime leaves
- Cilantro

- Bean sprouts
- Eggplant
- Green onions

Fruits:

- Mangoes
- Limes
- Pineapple
- Tamarind

Proteins:

- Chicken breasts
- Shrimp
- Tofu

Dairy and Eggs:

- Eggs
- Coconut milk

Herbs and Spices:

- Fresh ginger
- Garlic
- Galangal
- Thai bird's eye chilies
- Curry pastes (green, red, yellow)

Frozen Items:

- Frozen fish fillets
- Frozen dumplings

Miscellaneous:

- Rice noodles
- Sticky rice
- Palm sugar
- Fish sauce

Making Each Prompt Your Own: Customization Tips

- **Provide ChatGPT with Details About Your Menu**: Share specific dishes on your menu, and ChatGPT can adjust the checklist to focus on the key ingredients needed for those items. For instance, if you offer papaya salad, it can include green papaya and dried shrimp.
- **Request Quantities or Reorder Levels**: Specify that you'd like the checklist to include approximate quantities or notes like "reorder when stock is below 2 kg." ChatGPT can create a more detailed, actionable inventory.
- **Format the Checklist to Suit Your Workflow**: Ask ChatGPT to format the checklist for specific uses, such as creating printable templates, digital versions for Excel, or input-ready files for inventory apps.
- **Bonus Chapter:** Check out the Bonus Chapter to learn more about using this prompt

By providing details such as your restaurant name, menu highlights, and storage needs, ChatGPT can create an inventory checklist that is fully aligned with your operations and preferences.

Prompt 2: Reminder to Order Specific Ingredients for the Week

Prompt: *"Create a reminder to order essential ingredients for the week tailored to a Thai restaurant."*

Output from ChatGPT

"Reminder: Order Essential Ingredients for the Week (Thai Restaurant)"

📅 **Order Date**: [Insert Date]
🕒 **Deadline**: [Insert Time]
Must-Have Ingredients:
Rice & Noodles:

- Jasmine rice (bulk)
- Pad Thai noodles
- Glass noodles

Herbs & Spices:

- Lemongrass (fresh)

- Kaffir lime leaves
- Thai basil
- Cilantro
- Galangal root
- Thai bird's eye chilies
- Curry paste (red, green, yellow)
- Tamarind paste

Sauces & Condiments:

- Fish sauce
- Soy sauce (light & dark)
- Oyster sauce
- Palm sugar
- Coconut milk

Proteins:

- Chicken (whole or breasts/thighs)
- Pork (belly, ground, or tenderloin)
- Beef (stir-fry cuts)
- Shrimp (fresh or frozen)
- Tofu (firm)

Vegetables:

- Bean sprouts
- Bamboo shoots
- Thai eggplant
- Bell peppers (red, yellow, green)
- Mushrooms (shiitake, straw, or button)

- Snow peas

Other:

- Peanuts (crushed for garnish)
- Lime (fresh)
- Coconut (grated or shredded)
- Thai chili paste (Nam Prik Pao)

Special Orders (If Applicable):

- Thai desserts ingredients (sticky rice, mango, etc.)
- Specialty herbs (if needed)
- Extra stock of specialty items (e.g., tamarind, kaffir lime)

Making Each Prompt Your Own: Customization Tips

- **Provide ChatGPT with Your Weekly Needs**: Share the exact quantities or frequency of orders for your restaurant. For example, let ChatGPT know you need "10 kg of jasmine rice every week" or "shrimp twice a month." It can then generate more precise reminders.
- **Add Supplier Information**: Include details like supplier names, phone numbers, or websites, and ask ChatGPT to integrate them into the reminder. For instance, it could generate, "Order tofu from [Supplier Name] by 3 PM."
- **Incorporate Menu-Specific Ingredients**: If your menu features signature or seasonal items, let ChatGPT adjust the list to prioritize those ingredients. For example, if you're promoting mango sticky

rice, it can emphasize mangoes and sticky rice in bold.

- **Include Stock Tracking Notes**: Request a format where ChatGPT adds stock levels, like "Current: 5 kg jasmine rice, reorder at 2 kg," to help you identify low inventory items immediately.
- **Request Alerts for Specialty Ingredients**: Let ChatGPT know about hard-to-source or time-sensitive items. It can highlight them in the reminder with notes like, "Check specialty herb availability early this week."

Prompt 3: Weekly Ingredient Waste Reduction Email

Prompt: *"Write a weekly email to kitchen staff at a Thai restaurant, highlighting ingredients with high usage or waste from the previous week and suggesting ways to improve inventory management."*

Output from ChatGPT

"**Subject:** Weekly Ingredient Waste Review & Improvement Suggestions – [Restaurant Name]"

Dear Team,

Here's the ingredient waste review for the past week at [Restaurant Name], along with some strategies to help us reduce waste moving forward:

High Usage Ingredients

- **Thai basil**: Used heavily in dishes like Pad Graprow and Green Curry. We still have enough stock for this week, but let's monitor portions to avoid overuse.
- **Shrimp**: A large amount was used last week, especially in dishes like Tom Yum Soup and Pad Thai. Be mindful of portion control to prevent waste.
- **Coconut milk**: Consistently used in curries and soups, but we've had some leftover that may go unused this week. Let's incorporate it into additional dishes like Massaman Curry.

Waste Trends

- **Kaffir lime leaves**: A portion of last week's stock went unused and had to be discarded. Let's try to incorporate these more frequently into dishes like Tom Kha Gai or stir-fries.
- **Green onions**: Slightly more waste than usual. Consider trimming the stems and using them for garnishing or adding to broths to avoid discarding.

Suggestions for Improvement

- **Better portion control**: Review prep recipes, especially for dishes like Pad Thai and Green Curry, to avoid over-portioning ingredients.
- **Repurpose leftovers**: Use unused ingredients such as Thai basil or kaffir lime leaves in other menu items to maximize freshness.
- **Storage improvements**: Ensure proper sealing and refrigeration for perishable items like shrimp and coconut milk to extend shelf life.

Please keep these points in mind when preparing for the week ahead and let me know if you need any further guidance or supplies to reduce waste.

Thanks,

[Your Name]

Making Each Prompt Your Own: Customization Tips

- **Personalize the Waste Review Based on Your Menu**: ChatGPT can help tailor the ingredient list based on your restaurant's specific dishes. For example, if you specialize in Thai salads, you can ask ChatGPT to include higher-usage ingredients like Thai mint, tomatoes, or long beans. Providing more details about your menu will enable ChatGPT to generate a more accurate waste report.
- **Track Your Waste Patterns**: Share your restaurant's waste data with ChatGPT, such as waste percentages or inventory trends. This way, it can create detailed weekly reports reflecting specific ingredient waste issues like, "Last week, we used 10 kg of shrimp, but 2 kg was discarded due to spoilage." You can then adjust the reminders accordingly.
- **Use Specific Portion Guidelines**: Let ChatGPT know how much of each ingredient is typically used in each dish. For instance, you might specify that each serving of Pad Thai uses "30g of Bean Shoots" or "4 shrimps." With this information, ChatGPT can suggest more accurate portion control measures, ensuring your kitchen staff has a clear understanding of how much to use.
- **Add Suggestions for New Menu Items**: If you want to reduce waste, ChatGPT can help suggest menu items that utilize your excess ingredients. For example, "We have a surplus of kaffir lime

leaves. Consider adding a new stir-fry with kaffir lime for the next special."

- **Include Staff Responsibilities**: Ask ChatGPT to address specific staff members by name or role to take on waste-reduction tasks. For example, you might assign a sous chef to monitor shrimp portions or ask the prep cook to repurpose basil in the soups. ChatGPT can draft these personalized tasks directly into the message for clearer team communication.
- **Set Up Regular Reminders for Waste Reviews**: You can work with ChatGPT to create a recurring reminder for weekly ingredient waste audits. This can be automated and sent to your kitchen staff every week at a designated time, ensuring the team is always prepared to minimize waste at the start of each week.

By providing ChatGPT with accurate, restaurant-specific information, you'll be able to receive more tailored and actionable waste-reduction strategies for your Thai restaurant.

Staff Communication and Scheduling

Smooth communication and effective scheduling are essential for a harmonious and efficient restaurant environment. ChatGPT can assist in creating staff announcements, scheduling shifts, and ensuring everyone is on the same page.

Prompt 4: Staff Announcement for a Menu Update

Prompt: *"Write a staff announcement for a menu update."*

Output from ChatGPT

Staff Announcement: Menu Update!

Dear Team,

We are excited to announce that we've made some delicious updates to our menu! After carefully considering feedback from our guests and exploring some fresh ideas, we've introduced new items that we believe will enhance the dining experience at [Restaurant Name]. These updates will be rolling out [insert date], so here's a quick rundown of the changes:

New Dishes

- **[Dish 1 Name]** – [Brief description]
- **[Dish 2 Name]** – [Brief description]
- **[Dish 3 Name]** – [Brief description]

Updated/Improved Recipes

- **[Dish 1 Name]** – [What's changed or improved]
- **[Dish 2 Name]** – [What's changed or improved]

Items Discontinued

- **[Dish 1 Name]**
- **[Dish 2 Name]**

Important Notes

- Please take a moment to familiarize yourself with the new items and updates on the menu.
- Our servers and bartenders will be updated on any changes in specials, ingredients, and pricing.
- If you have any questions about the new menu or need help with recommendations for guests, don't hesitate to ask the kitchen or management team.

We're excited to see how our guests respond to the changes and are confident these new dishes will become new favorites! As always, thank you for your continued hard work and dedication. Let's make this menu rollout a huge success!

Warm regards,
[Your Name]
[Your Position]
[Restaurant Name]

Making Each Prompt Your Own: Customization Tips

- **Highlight Specific Allergen or Dietary Information**: When updating your menu, it's essential to provide your team with information on any new allergens or dietary restrictions related to the dishes. You can ask ChatGPT to include details like "This dish contains peanuts" or "Vegan options available" to ensure your staff can provide accurate recommendations to customers with dietary concerns.

- **Include Staff Training and Tasting Sessions**: If you're offering training or tasting sessions to help your team get familiar with the new menu, be sure to mention it in your announcement. ChatGPT can craft a section such as: "Join us for a tasting session on [date] where we'll go through the new dishes and provide tasting notes."
- **Include Pricing and Portion Changes**: If your menu update includes price changes or adjusted portion sizes, make sure to address this in the announcement. You could say, "New menu items will be priced as follows..." and list the price next to each dish. For any price adjustments to existing dishes, ask ChatGPT to include this detail so your team is prepared to communicate these changes to guests confidently.
- **Clarify Preparation or Cooking Techniques**: If the new dishes or improved recipes involve changes to the way your kitchen prepares or cooks food, ChatGPT can specify the details for your kitchen team. For example, "The new [Dish Name] will require the addition of [specific ingredient] during the cooking process" or "Please note that this dish needs to be cooked at a higher heat to maintain the right texture." These technical notes ensure consistency and reduce errors in food preparation.
- **Add Incentives or Motivational Elements**: To energize your team, ask ChatGPT to add motivational language, such as "We're excited to introduce these new dishes, and we know they'll be a hit with our guests!" or "Thank you for your hard work, and let's make sure these menu changes go smoothly!" This can help rally the staff around the new menu and build excitement for the rollout.
- **Include a Digital or Print Menu Link**: If you're sharing a digital menu or have printed materials ready for the team to review, ask ChatGPT to include a link or instructions on where to find them.

For example: "You can view the updated menu online here [insert link], or pick up a printed copy in the staff room."

By providing ChatGPT with your specific menu updates, staff training plans, and tone preferences, you can generate a staff announcement that keeps everyone informed and excited about the changes.

Prompt 5: Weekly Staff Schedule for Kitchen and Front-of-House

"See Bonus Chapter for an Expanded Version"

Prompt: *"Create a weekly schedule for kitchen and front-of-house staff."*

Output from ChatGPT

Weekly Staff Schedule – [Restaurant Name]

Monday – Friday:

Kitchen:

- **Chef [Name]** – 9 AM – 5 PM
- **Sous Chef [Name]** – 10 AM – 6 PM
- **Line Cook [Name]** – 11 AM – 7 PM
- **Prep Cook [Name]** – 9 AM – 3 PM

Front-of-House:

- **Host [Name]** – 10 AM – 6 PM
- **Server [Name]** – 11 AM – 9 PM
- **Bartender [Name]** – 4 PM – 10 PM
- **Busser [Name]** – 5 PM – 11 PM

Saturday & Sunday (Peak Hours):
Kitchen:

- **Chef [Name]** – 10 AM – 8 PM
- **Sous Chef [Name]** – 11 AM – 9 PM
- **Line Cook [Name]** – 12 PM – 10 PM
- **Prep Cook [Name]** – 9 AM – 5 PM

Front-of-House:

- **Host [Name]** – 11 AM – 5 PM
- **Server [Name]** – 12 PM – 10 PM
- **Bartender [Name]** – 2 PM – 10 PM

Making Each Prompt Your Own: Customization Tips

- **Staff Availability & Preferences:** Tailor shifts to accommodate staff schedules, such as preferred shifts or specific role requests (e.g., "Server [Name] prefers mornings").
- **Special Events & Holidays:** Adjust for busy periods by adding extra staff or modifying shifts to cover high-demand times (e.g., weekends, holidays, or special events).
- **Peak Hours & Flexibility:** Customize the schedule for peak hours by adding more coverage (e.g., "Increase kitchen staff during dinner

rush") and allow flexibility for last-minute changes.

- **Leadership & Role Assignments:** Clearly define leadership roles and responsibilities (e.g., "Manager oversees kitchen during dinner") and ensure roles like servers or chefs are assigned to specific tasks.
- **Breaks & Shift Overlaps:** Incorporate break times and schedule staff overlap to ensure smooth transitions between shifts (e.g., "Server [Name] overlaps with host from 4 PM to 6 PM").
- **Bonus Chapter:** Be sure to check out the Bonus Chapter for more insights on how you can expand and personalize this prompt even further.

These tips will help tailor the schedule to your restaurant's unique needs and staff dynamics.

Prompt 6: Staff Shift Change Notification

Prompt: *"Create a staff shift change notification for a team member switching shifts."*

Output from ChatGPT

"Subject: Shift Change Notification – [Restaurant Name]

Dear [Staff Member],

This is to inform you that your shift has been changed for [Date]. You'll now be working [new shift time].

Please confirm your availability as soon as possible, and let me know if you have any questions.

Thank you for your flexibility!

Best regards,

[Your Name]"

Making Each Prompt Your Own: Customization Tips

- **Tone**: Adjust the tone to match your restaurant's culture—formal for fine dining, casual for fast-casual spots, or friendly and approachable for a more relaxed atmosphere.
- **Clear Shift Details**: Include exact times, roles, and responsibilities for the shift change to avoid confusion (e.g., "You'll now be covering prep duties from 10 AM to 4 PM").
- **Advance Notice**: Provide ample notice for shift changes, especially for last-minute adjustments, so the employee has time to confirm their availability.
- **Additional Information**: Mention any special tasks or events tied to the shift change, such as new menu items to promote or VIP customers to expect.

Follow-up Communication: Offer an option for the staff member to reach out if there are concerns or issues with the change (e.g., "Please feel free to contact me if you have any conflicts").

Training and Development

Ongoing staff training is essential for providing excellent service and maintaining consistency. ChatGPT can help you create training materials, quizzes, and onboarding documents to help your team stay updated on best practices and new procedures.

Prompt 7: Training Manual for New Kitchen Staff

Prompt: *"Write a training manual for new kitchen staff on food safety protocols."*

Output from ChatGPT

New Kitchen Staff Food Safety Manual

Welcome to [Restaurant Name]! We're dedicated to providing a safe and clean kitchen environment for both our staff and customers. Here are the key food safety protocols:

Personal Hygiene

- Wash hands thoroughly before handling food and after using the restroom.

- Wear gloves when handling ready-to-eat foods.
- Keep nails trimmed and avoid wearing jewelry.
- Hair must be tied back and covered at all times.

Temperature Control

- Cold foods should be stored at or below 40°F and hot foods above 140°F.
- Use food thermometers to check the internal temperature.
- Always check that food is properly cooked before serving.

Cross-Contamination Prevention

- Separate raw meat, seafood, and vegetables during preparation.
- Clean and sanitize all work surfaces frequently.

Making Each Prompt Your Own: Customization Tips

- **ChatGPT Customization**: Ask ChatGPT to adapt the food safety manual to your restaurant's specific needs, like addressing unique ingredients or menu items.
- **ChatGPT-Generated Resources**: Request that ChatGPT provide additional materials, such as video links, illustrations, or step-by-step procedures, to reinforce training.
- **Tailored Protocols**: Use ChatGPT to adjust the safety protocols to align with local regulations and best practices, ensuring your staff follows the most up-to-date guidelines.
- **Kitchen-Specific Equipment**: Have ChatGPT incorporate your kitchen's unique equipment and tools (e.g., specialized thermome-

ters or storage units) into the safety instructions.

Cultural Relevance: If your restaurant has specific culinary techniques (like raw fish preparation), ask ChatGPT to emphasize those aspects within the manual to ensure staff understand the special precautions.

Prompt 8: Staff Quiz on Food Safety Best Practices

"See Bonus Chapter for an Expanded Version"

Prompt: *"Generate a staff quiz on food safety best practices for kitchen staff."*

Output from ChatGPT

Food Safety Quiz for Kitchen Staff:

What should you do before handling food?

a) Wash your hands thoroughly

b) Only wear gloves

c) Wait until after touching equipment

What is the proper internal temperature for cooked poultry?

a) 145°F

b) 165°F

c) 200°F

How should raw meat be stored in the refrigerator?

a) On the top shelf

b) On the bottom shelf to prevent cross-contamination

c) Mixed with other foods

Making Each Prompt Your Own: Customization Tips

- **ChatGPT Personalization**: Ask ChatGPT to create quiz questions specific to your menu items, such as questions about handling raw seafood, gluten-free protocols, or unique ingredients like sushi rice.
- **Customize Difficulty**: Use ChatGPT to adjust the quiz's difficulty level, catering to both new hires and experienced staff, by including both basic and advanced food safety topics.
- **Focus on Key Areas**: Have ChatGPT include questions that address particular concerns in your kitchen, like allergy management or cross-contamination between different food groups.
- **Tailor Feedback**: Request that ChatGPT generates a feedback section for the quiz to help guide staff on improving food safety practices based on their answers.
- **Question Quantity**: Specify the number of questions you'd like ChatGPT to include in the quiz and whether you'd like multiple-choice, true/false, or short-answer formats for variety.
- **Bonus Chapter:** For additional guidance on expanding this prompt, refer to the Bonus Chapter, where we dive deeper into the details.

Prompt 9: Creating a New Staff Training Course

Prompt: *"Write a training outline for a new customer service course for front-of-house staff."*

Output from ChatGPT

Customer Service Excellence Training Outline:

Introduction to Customer Service

- The importance of exceptional service
- Key characteristics of a great server

Handling Customer Complaints

- Active listening and empathy
- How to diffuse difficult situations

Upselling and Suggestive Selling

- How to recommend menu items naturally
- Providing a memorable dining experience

Effective Communication

- Using positive language
- Ensuring smooth communication with kitchen staff

Making Each Prompt Your Own: Customization Tips

- **ChatGPT to Tailor Content**: Ask ChatGPT to customize the training outline based on the specific atmosphere of your restaurant, whether it's fine dining, casual, or fast food. This ensures the customer service approach aligns with your brand.
- **Incorporate Specific Scenarios**: Use ChatGPT to include examples or role-playing exercises that reflect your restaurant's common customer interactions, from handling dietary restrictions to managing large party bookings.
- **Integrate Your Restaurant's Policies**: Request ChatGPT to weave in your restaurant's policies on customer service, such as how to handle complaints, discounts, or loyalty program queries.
- **Customize Training Length**: Specify the desired length of the course and ask ChatGPT to organize content in manageable modules, focusing on key areas like greetings, problem-solving, and closing the sale.
- **Enhance Engagement**: Have ChatGPT suggest ways to keep the course interactive, including quizzes or group discussions, to make the training more engaging for your staff.
- **Write the course:** Once you're satisfied with the training outline provided by ChatGPT, take the next step and ask it to create the full customer service course tailored for your front-of-house staff, ensuring it follows the outline it developed. This is where the real magic begins—you'll be amazed by the comprehensive and thoughtful course content ChatGPT can deliver!

Prompt 10: Staff Onboarding Checklist for New Hires

Prompt: *"Create an onboarding checklist for new kitchen and front-of-house staff."*

Output from ChatGPT

"New Staff Onboarding Checklist

Introduction to the Team

- Meet your manager and coworkers
- Tour of the restaurant

Workplace Policies

- Review the employee handbook
- Discuss dress code, break times, and shift expectations

Role-Specific Training

- Kitchen staff: Cooking methods, equipment handling
- Front-of-house staff: Customer interaction, POS system training

Health and Safety Procedures

- Kitchen hygiene, handling allergens
- Emergency procedures

Making Each Prompt Your Own: Customization Tips

- **Use ChatGPT for Role-Specific Customization**: Customize the checklist for different roles by having ChatGPT generate specific tasks for kitchen vs. front-of-house staff, such as unique kitchen procedures or POS system training for servers.
- **Incorporate Restaurant Culture**: Ask ChatGPT to help incorporate your restaurant's values and culture into the onboarding process. For example, focus on hospitality if you're a high-end restaurant or teamwork if you're a casual, fast-paced establishment.
- **Add Detailed Health & Safety Procedures**: Use ChatGPT to tailor safety protocols to your kitchen environment, such as food handling techniques specific to your menu, allergens, or health and sanitation practices.
- **Personalize Employee Integration**: Request ChatGPT to suggest ways to introduce new hires to the team, such as mentoring programs or informal meet-and-greets, to help integrate them into the work culture.
- **Track Training Progress**: Use ChatGPT to create a follow-up system, ensuring that new hires complete essential training tasks on time, such as sign-offs on safety protocols or completing specific role-related tasks.

Conclusion

Before we dive into financial management, **I'd love to hear your thoughts!** If you've found the insights and strategies in this book helpful so far, leaving a **quick review** not only helps other restaurant owners discover these tools but also allows me to keep improving and providing valuable content. **Your feedback makes a huge difference!**

Operational efficiency forms the backbone of a well-run restaurant. With ChatGPT, you can streamline tasks, improve communication, and keep your team organized and motivated. These prompts provide a starting point, but by customizing them, you'll create systems that work seamlessly for your unique operation.

Now that your operations are optimized, let's explore how ChatGPT can help you tackle **financial management** in the next chapter. From budgeting to expense tracking, discover how to take control of your restaurant's finances with ease.

7

Financial and Business Strategy

In the fast-paced restaurant industry, maintaining financial stability and profitability is crucial for long-term success. It's not just about serving great food—it's also about managing costs, optimizing revenue, and ensuring a consistent customer experience. In this chapter, we'll explore how ChatGPT can be used to help with cost analysis, budgeting, profitability strategies, and implementing technologies. We'll also provide examples and practical prompts, along with tips on how to customize them for your specific restaurant needs.

Using ChatGPT for Cost Analysis and Budgeting

Effective financial management begins with analyzing costs and creating a realistic budget. ChatGPT can help you assess ingredient costs, prepare budget-friendly strategies for monthly operations, and determine the appropriate pricing for your menu items. Let's dive into some examples of how you can utilize this tool to make more informed financial decisions.

Prompt 1: Pricing a new menu item based on the cost of ingredients

"See Bonus Chapter for an Expanded Version"

Prompt: *"Analyze the cost of ingredients for a new menu item (Grilled Thai Pork Skewers) and suggest a reasonable price point."*

Output from ChatGPT

To calculate the cost and suggest a price for your new menu item, let's break down the cost of each ingredient involved:

- **Pork (per 3 skewers)**: $2.50
- **Marinade (fish sauce, garlic, brown sugar, lime juice, herbs)**: $0.80
- **Rice (per serving)**: $0.50
- **Garnishes (cilantro, peanuts)**: $0.30

Total Food Cost (per 3 skewers): $4.10

To determine the price point, we'll calculate the **food cost percentage**. A standard target food cost percentage for most restaurants is

30-35%, but it can vary based on your establishment's positioning (e.g., casual dining versus fine dining). For this example, we'll aim for a 30% food cost.

Suggested Menu Price = Total Food Cost ÷ Desired Food Cost Percentage

Suggested Menu Price = $4.10 ÷ 0.30 = **$13.67** (rounded to $13.95)

Thus, the suggested price for the Grilled Thai Pork Skewers would be **$13.95** to maintain a 30% food cost.

Making Each Prompt Your Own: Customization Tips

- **Refine Cost Calculations with ChatGPT:** Input your exact ingredient costs, portion sizes, and supplier prices, and let ChatGPT recalculate the food cost percentage for your restaurant.
- **Adjust Pricing Strategy:** Ask ChatGPT to recommend a price based on your business model—fine dining may allow for a higher markup, while a casual eatery might aim for affordability.
- **Factor in Overhead & Profit Margins:** Use ChatGPT to include additional costs like labor, rent, and utilities in your pricing formula to ensure profitability.
- **Compare with Competitors:** Have ChatGPT analyze competitor pricing in your area to ensure your price is competitive while maintaining profitability.
- **Experiment with Upsells & Bundles:** Ask ChatGPT to suggest ways to increase revenue, such as offering the dish as part of a combo meal, adding premium options, or pairing it with high-margin beverages.
- **Bonus Chapter:** For a more detailed guide on customizing this prompt to fit your restaurant's specific needs, see the bonus chapter.

Prompt 2: Creating a monthly budget for the restaurant operations

Prompt: *"Create a budget for monthly restaurant operations, including labor, food, and utilities."*

Output from ChatGPT

Here's an example of how a monthly operations budget for a mid-sized restaurant might look:

1. Labor Costs:

Labor is one of the largest expenses in a restaurant. Labor costs include wages, benefits, and payroll taxes for both front-of-house and kitchen staff.

- **Management Salaries** (e.g., General Manager, Executive Chef): $8,000
- **Front-of-House Staff** (servers, bartenders, hosts): $12,000
- **Kitchen Staff** (line cooks, prep cooks, dishwashers): $15,000
- **Payroll Taxes and Benefits**: $3,000

Total Labor Costs: $38,000

2. Food & Beverage Costs:

Food and beverage costs can vary greatly depending on the menu offerings and suppliers. The typical food cost percentage in the restaurant industry is between 25% and 35%, depending on your menu and pricing strategy.

- **Food Purchases** (ingredients for dishes): $25,000
- **Beverages** (alcohol, non-alcoholic drinks): $7,000

Total Food & Beverage Costs: $32,000

3. Utilities:

Utilities like water, gas, electricity, and phone services are crucial to keeping your restaurant operational.

- **Electricity**: $2,000
- **Water**: $800
- **Gas**: $1,500
- **Internet and Phone**: $300

Total Utilities Costs: $4,600

4. Rent & Other Fixed Costs:

Rent is a major fixed cost, along with other recurring expenses like insurance, marketing, and POS systems.

- **Rent**: $6,000
- **Insurance**: $500
- **Marketing**: $2,000
- **POS System Subscription**: $200

Total Fixed Costs: $8,700

Total Monthly Expenses: $83,300

To calculate your revenue target, you need to understand the desired profit margin. Most restaurants aim for a 25-30% profit margin after covering all expenses. Let's say your desired profit margin is 30%. Your target revenue would be:

Target Revenue = Total Expenses ÷ (1 - Desired Profit Margin)

Target Revenue = $83,300 ÷ (1 - 0.30) = **$119,000**

Thus, your monthly revenue target should be approximately **$119,000**.

Making Each Prompt Your Own: Customization Tips

- **Tailor Expense Categories with ChatGPT**: Provide ChatGPT with details on your specific costs (e.g., labor, rent, or marketing), and it will generate a budget breakdown customized for your restaurant.
- **Adjust Based on Restaurant Type**: Whether you're a food truck, fine dining, or quick-service restaurant, ChatGPT can help refine cost estimates and ideal profit margins accordingly.
- **Factor in Seasonal Variations**: Ask ChatGPT to adjust the budget for seasonal fluctuations in food prices, staffing needs, and utility costs.
- **Optimize for Profitability**: Use ChatGPT to explore cost-cutting strategies, such as supplier negotiations, energy-efficient equipment, or labor scheduling improvements.
- **Set Revenue Goals with Precision**: ChatGPT can calculate your ideal revenue target based on your expenses and desired profit margin, helping you set realistic financial goals.

Prompt 3: Calculating the cost of a large catering order

Prompt: *"Calculate the cost of a catering order for 50 people, including Thai appetizers, main courses, and desserts. Provide a breakdown of costs and suggest a pricing structure for the order."*

Output from ChatGPT

Let's break down the cost of the catering order, focusing on traditional Thai appetizers, main courses, and desserts. We'll also factor in the unique ingredients and typical portion sizes for a Thai restaurant.

1. Thai Appetizers (per person):

For appetizers, let's offer a selection of traditional Thai dishes such as spring rolls, satay skewers, and Thai fish cakes.

- **Thai Spring Rolls (serves 4 people)**: $4.00
- **Chicken Satay Skewers (serves 4 people)**: $6.00
- **Thai Fish Cakes (serves 4 people)**: $5.00

Cost per person (appetizers) = ($4 + $6 + $5) ÷ 4 = **$3.75**

Total cost for 50 people = 50 x $3.75 = **$187.50**

2. Thai Main Course (per person):

For the main course, let's assume a mix of popular Thai dishes such as Pad Thai, Green Curry with chicken, and a vegetarian stir-fry.

- **Pad Thai (serves 1 person)**: $7.00

- **Green Curry with Chicken (serves 1 person)**: $8.00
- **Vegetarian Stir-fry (serves 1 person)**: $6.00

Cost per person (main course) = $7.00 + $8.00 + $6.00 = **$21.00**
Total cost for 50 people = 50 x $21.00 = **$1,050.00**

3. Thai Desserts (per person):
For dessert, let's offer traditional Thai desserts such as sticky rice with mango and coconut custard.

- **Sticky Rice with Mango (serves 1 person)**: $3.00
- **Coconut Custard (serves 1 person)**: $2.50

Cost per person (dessert) = $3.00 + $2.50 = **$5.50**
Total cost for 50 people = 50 x $5.50 = **$275.00**

Total Catering Order Cost:

- **Appetizers**: $187.50
- **Main Course**: $1,050.00
- **Desserts**: $275.00

Total Cost for Catering Order = $187.50 + $1,050.00 + $275.00 = **$1,512.50**

Now, let's determine a fair price for the catering order.

A standard target catering profit margin for Thai cuisine could range from 20-30%. For the sake of this example, we'll use a **25% profit margin**.

Suggested Catering Price = Total Catering Cost ÷ (1 - Desired Profit Margin)

Suggested Catering Price = $1,512.50 ÷ (1 - 0.25) = **$2,016.67**
Suggested Price for Catering Order = $2,015.00

Thus, the suggested price for this catering order for 50 people would be **$2,015.00**.

Making Each Prompt Your Own: Customization Tips

- **Customize the Menu with ChatGPT**: Provide ChatGPT with your restaurant's exact catering dishes, portion sizes, and ingredient costs to generate a precise cost breakdown.
- **Adjust for Portion Sizes & Ingredients**: ChatGPT can help you scale portion sizes for larger catering orders and factor in premium ingredients like authentic Thai herbs or specialty meats.
- **Set a Competitive Profit Margin**: Ask ChatGPT to adjust pricing based on your target profit margin, whether you're aiming for affordability or a premium catering experience.
- **Account for Additional Costs**: ChatGPT can calculate delivery fees, staffing costs, and other service charges to ensure your pricing covers all expenses.
- **Cater to Dietary Needs:** If you offer gluten-free, vegetarian, or allergy-friendly options, ChatGPT can adjust pricing and ingredient costs accordingly.

Profitability Strategies

Increasing your profit margins requires a strategic approach that balances cost-cutting with revenue optimization. ChatGPT can help you analyze your pricing structure, adjust menu items, and identify new ways to increase profitability while maintaining quality and customer satisfaction.

Prompt 4: Suggestions for increasing profit margins

"See Bonus Chapter for an Expanded Version"

Prompt: *"Generate suggestions for increasing profit margins without compromising quality."*

Output from ChatGPT

Here are some strategies to consider for increasing profit margins:

Menu Engineering:

- **Optimize High-Profit Items**: Identify dishes with high-profit margins and make them stand out on your menu. Consider adding creative descriptions, pairing them with complementary sides, or recommending them as specials.

- **Remove Low-Margin Items**: If certain items consistently underperform or have low margins, consider removing or revamping them.
- **Offer Upgrades and Add-Ons**: Offer customers the option to upgrade their meal (e.g., premium sides, larger portions, or additional toppings) for a slightly higher price.

Improve Portion Control:

- **Standardize Portions**: Ensure your kitchen staff is following consistent portion sizes for all dishes to avoid waste and improve cost control.
- **Smaller Plates**: Research shows that serving food on smaller plates can help reduce food waste while encouraging customers to finish their meals.

Labor Efficiency:

- **Optimize Scheduling**: Use data to understand peak dining hours and adjust your staff schedules accordingly. Avoid overstaffing during slow periods.
- **Cross-Train Staff**: Training employees to perform multiple roles can help reduce labor costs while maintaining service levels during peak hours.

Use Technology for Efficiency:

- **POS Systems**: Leverage your POS system to track inventory in real time, monitor food waste, and identify slow-moving items. This data can help you make informed decisions to reduce waste and streamline your operations.

- **Online Ordering & Delivery**: If your restaurant offers delivery, optimizing your online ordering system can help increase sales and profit margins.

Making Each Prompt Your Own: Customization Tips

- **Use ChatGPT for Menu Optimization**: Ask ChatGPT to analyze your menu, highlight high-margin items, and suggest pricing or upselling strategies tailored to your restaurant.
- **Fine-Tune Portion Control with ChatGPT**: Provide ChatGPT with portion size details, and it can generate guidelines for reducing waste without impacting customer satisfaction.
- **Enhance Labor Efficiency**: ChatGPT can help you create optimized staff schedules, cross-training plans, and labor cost analyses based on your restaurant's peak hours.
- **Leverage Technology for Cost Savings**: Ask ChatGPT how to integrate POS data, track inventory, and streamline online ordering to improve profit margins.
- **Tailor Strategies to Your Brand**: ChatGPT can customize cost-saving ideas based on your restaurant type—whether fast casual, fine dining, or family-style—ensuring they align with your target audience.
- **Bonus Chapter:** Don't forget to have a look at the Bonus Chapter for a closer look at customizing this prompt.

Prompt 5: Using ChatGPT to reduce food waste

Prompt: *"Create a list of 5 actionable tips to reduce food waste."*

Output from ChatGPT

1. Introduce daily specials to use surplus ingredients.
2. Train staff to track expiration dates and practice FIFO (First In, First Out).
3. Repurpose leftover ingredients for soups, stocks, or garnishes.
4. Offer smaller portion sizes or half orders for waste-conscious diners.
5. Partner with local charities to donate excess food.

Making Each Prompt Your Own: Customization Tips

- **Use ChatGPT for Waste Reduction Strategies**: Ask ChatGPT to analyze your menu and suggest creative ways to repurpose excess ingredients based on your cuisine.
- **Optimize Inventory Management**: ChatGPT can help you implement FIFO tracking methods, expiration date monitoring, and portion control strategies to minimize waste.
- **Customize Specials to Reduce Waste**: Provide ChatGPT with a list of surplus ingredients, and it can generate daily specials that align with your restaurant's theme.
- **Improve Portion Control & Customer Options**: ChatGPT can assist in designing flexible portion sizes, half orders, or

customizable dishes to reduce plate waste.

- **Find Donation & Sustainability Partners**: Use ChatGPT to identify local food donation programs, composting initiatives, or partnerships with farms to repurpose food scraps.

Prompt 6: Identifying Cost-Saving Opportunities in Ingredient Sourcing

Prompt: *"Suggest cost-saving strategies for sourcing high-quality ingredients without compromising food quality."*

Output from ChatGPT

Here are some cost-saving strategies to consider for sourcing high-quality ingredients:

- **Build Strong Relationships with Suppliers:** Negotiate long-term contracts with reliable suppliers to secure better pricing for bulk purchases. Consider working with local farmers or co-ops for seasonal discounts.
- **Buy Seasonal Ingredients:** Utilize ingredients that are in season, as they tend to be fresher, more abundant, and less expensive. Build your menu around what's readily available.
- **Reduce Food Waste:** Analyze purchasing trends and ensure you're ordering only what you need to reduce overstocking and spoilage. Consider using leftover ingredients creatively in specials or sides.

- **Evaluate Multiple Vendors:** Regularly review supplier contracts and compare prices to ensure you're getting the best deal. Avoid being tied to a single vendor when better deals might be available elsewhere.
- **Leverage Group Purchasing Organizations (GPOs):** If you're part of a restaurant group or association, use collective buying power to access discounts.
- **Focus on Versatile Ingredients:** Choose ingredients that can be used in multiple dishes across your menu to optimize inventory and reduce wastage.

Making Each Prompt Your Own: Customization Tips

- **Use ChatGPT for Supplier Negotiations**: Ask ChatGPT to draft emails or scripts for negotiating better pricing with suppliers, bulk discounts, or seasonal deals.
- **Find Cost-Effective Ingredient Alternatives**: ChatGPT can suggest substitutes for expensive ingredients that maintain the integrity of your dishes.
- **Optimize Seasonal & Bulk Purchasing**: Provide ChatGPT with your menu, and it can recommend seasonal ingredient swaps or bulk-buying strategies to save costs.
- **Evaluate Vendors & Contracts**: Use ChatGPT to compare supplier options, generate cost-analysis reports, and suggest alternative sourcing methods like co-ops or GPOs.
- **Minimize Waste & Maximize Efficiency**: ChatGPT can analyze your ingredient usage, suggest ways to repurpose leftovers, and streamline purchasing to prevent overstocking.
- **Cuisine-Specific Adjustments:** Depending on your restaurant's

theme, certain ingredients may require specialized sourcing. For example, an Italian restaurant might negotiate deals with olive oil importers.

- **Test Supplier Alternatives:** Consider trialing new vendors to gauge quality while assessing their pricing structure. Ensure the new supplier aligns with your standards before making permanent changes.
- **Feedback Loop:** Involve your kitchen team in tracking ingredient quality and costs over time to identify areas for improvement or negotiation.

Prompt 7: Increasing Beverage Profit Margins

Prompt: *"Provide strategies for increasing profit margins on beverages, including cocktails, wine, and non-alcoholic drinks."*

Output from ChatGPT

Here are some strategies to increase profit margins on beverages:

- **Create Signature Drinks:** Design unique cocktails or mocktails that align with your brand and are exclusive to your restaurant. These can command premium pricing.
- **Focus on Wine Pairing Menus:** Offer curated wine pairings with your menu, priced as a bundle. Highlighting the pairing experience adds value and increases sales.

- **Upsell Premium Options:** Train servers to recommend higher-quality wines, spirits, or mixers, which often have higher profit margins.
- **Optimize Portion Sizes:** Use precise pour controls for alcoholic beverages to reduce waste and ensure consistent serving sizes. Consider implementing smaller portions for high-end drinks at a premium price.
- **Promote Non-Alcoholic Beverages:** Create visually appealing non-alcoholic options like mocktails, fresh juices, or specialty sodas that cater to a broader audience.
- **Host Beverage-Centric Events:** Organize events such as wine tastings, cocktail workshops, or seasonal drink launches to attract customers and generate additional revenue.
- **Streamline Bar Inventory:** Regularly analyze beverage sales and remove slow-moving inventory. Focus on stocking high-demand, profitable items.

Making Each Prompt Your Own: Customization Tips

- **Use ChatGPT for Custom Cocktail & Mocktail Creation**: Provide ChatGPT with your restaurant's theme, and it can generate unique, high-margin drink recipes tailored to your menu.
- **Train Staff with ChatGPT's Sales Scripts**: Ask ChatGPT to create upselling scripts or role-play scenarios to help staff confidently recommend premium beverages.
- **Analyze Beverage Costs & Pricing**: Use ChatGPT to calculate ideal pricing for cocktails, wine, and soft drinks based on ingredient costs and target profit margins.
- **Design Seasonal Drink Menus**: Let ChatGPT suggest limited-

time beverages based on seasonal ingredients, customer trends, or upcoming holidays to boost interest and sales.

- **Optimize Bar Inventory Management**: ChatGPT can help analyze beverage sales data, suggest slow-moving items to phase out, and recommend high-profit alternatives.
- **Customer Preferences:** Use data from your POS system to understand popular beverage choices and adjust pricing or availability to maximize profit.
- **Brand Synergy:** Ensure that your beverage offerings complement your restaurant's menu and ambiance. For example, a fine-dining establishment might prioritize premium wine pairings, while a casual spot might focus on craft beers or creative cocktails.

Technology-Driven Solutions

Technology simplifies operations, reduces costs, and boosts efficiency. Tools like POS systems and inventory management help minimize waste and optimize staffing. Embracing these solutions can streamline processes and improve profitability.

Prompt 8: Optimizing Online Reservations and Waitlists

Prompt: *"Suggest ways to use technology to improve reservations and waitlist management for my restaurant."*

Output from ChatGPT

Here's how technology can improve your reservations and waitlist processes:

- **Reservation Platforms:** Use tools like OpenTable, Resy, or Yelp Reservations to allow customers to book tables online with real-time availability.
- **Waitlist Apps:** Implement apps like Waitlist Me or Nowait to enable customers to join your waitlist remotely and receive text updates about their table status.
- **Automated Table Management:** Use software to optimize table turnover rates, ensuring maximum occupancy without overbooking.
- **Data Collection:** Track reservation patterns and customer preferences to personalize service, such as remembering special occasions or seating preferences.
- **Dynamic Pricing:** Adjust reservation pricing for high-demand times, such as offering premium pricing for peak slots or discounts for slower periods.
- **No-Show Prevention:** Send automated text or email reminders to customers before their reservations to reduce no-shows.

Making Each Prompt Your Own: Customization Tips

- **Choose the Right Tech for Your Restaurant Size**: For small restaurants, opt for simple, cost-effective waitlist apps like Waitlist Me. Larger venues may benefit from advanced platforms like OpenTable or SevenRooms.
- **Use ChatGPT to Draft Reservation Policies**: Ask ChatGPT to generate reservation confirmation messages, cancellation policies, or scripts for handling overbookings.
- **Personalize the Customer Experience**: Use ChatGPT to analyze reservation data and suggest ways to offer VIP treatment, such as remembering guests' favorite tables or special occasions.
- **Implement Smart Table Turnover Strategies**: Have ChatGPT suggest seating strategies to maximize efficiency, like optimizing party sizes for peak dining hours.
- **Reduce No-Shows with Automated Messaging**: Use ChatGPT to craft friendly, brand-aligned reservation reminders via text or email to keep guests informed and engaged.
- **Restaurant Size:** Smaller establishments might benefit from simpler, budget-friendly apps, while larger venues could invest in comprehensive reservation systems with analytics capabilities.
- **Brand Voice:** Ensure any customer-facing communications (e.g., text reminders) align with your restaurant's tone and branding.
- **Requesting More Information:** If you're unfamiliar with any software or app recommendations from ChatGPT, like 'OpenTable' or 'Waitlist Me', simply ask for more details. ChatGPT can provide an overview, highlight key features, and explain how they can benefit your restaurant.

Prompt 9: Enhancing Customer Engagement with Digital Tools

Prompt: *"What are some digital tools I can use to engage customers and encourage repeat visits?"*

Output from ChatGPT

Here are several digital tools to enhance customer engagement:

- **Loyalty Programs:** Implement digital loyalty apps like FiveStars or Toast Loyalty, allowing customers to earn rewards for frequent visits or purchases.
- **Email Marketing Platforms:** Use services like Mailchimp or Constant Contact to send newsletters with special offers, menu updates, and exclusive events.
- **Social Media Integration:** Use platforms like Instagram or Facebook to share visually appealing content, including new dishes, customer stories, or promotions.
- **Mobile Apps:** Develop a custom app for your restaurant that allows customers to make reservations, order takeout, or join loyalty programs seamlessly.
- **Feedback Collection:** Use tools like Google Forms or SurveyMonkey to gather customer feedback, providing insights into areas for improvement.
- **Push Notifications:** Send targeted push notifications about specials or discounts through your app or loyalty program to drive customer visits.

Making Each Prompt Your Own: Customization Tips

- **Leverage Loyalty Programs Based on Your Restaurant's Vibe**: If you run a casual or family-style restaurant, opt for simple, easy-to-use apps like FiveStars that offer quick rewards. For upscale establishments, consider more sophisticated programs that offer exclusive perks for repeat guests, like Toast Loyalty.
- **Craft Targeted Email Campaigns**: Ask ChatGPT to generate engaging email content for special events or new menu items. Personalize the emails to highlight the customer's previous visits or favorite dishes.
- **Maximize Social Media Engagement**: Use ChatGPT to draft creative Instagram or Facebook posts that highlight your restaurant's signature dishes, customer stories, and seasonal promotions. Use hashtags to improve visibility and engagement.
- **Develop a Custom Mobile App**: If your restaurant has a lot of takeout or delivery orders, consider a dedicated app where customers can easily reorder their favorite dishes or access loyalty rewards.
- **Encourage Feedback and Improve**: Use ChatGPT to craft personalized surveys to understand what customers love and what could be better, then act on that feedback to improve the experience.
- **Seasonal Push Notifications**: Use digital tools to send time-sensitive notifications for seasonal dishes or limited-time offers, especially around holidays or local events. ChatGPT can help you craft compelling, seasonal messaging that aligns with your restaurant's branding.
- **Focus on Demographics:** Tailor your digital strategy to your customer base—social media for younger audiences or loyalty programs for repeat diners.

- **Requesting More Information:** If you're not familiar with any of the software or app recommendations from ChatGPT, such as 'FiveStars' or 'Mailchimp', you can ask ChatGPT for more information. It can provide a brief overview, key features, and how they might benefit your restaurant.

Prompt 10: Automating Customer Feedback and Review Management

Prompt: *"How can I use technology to automate customer feedback collection and manage online reviews?"*

Output from ChatGPT

Here are some ways to automate feedback collection and streamline review management:

- **Automated Feedback Requests:** Use tools like Podium or Birdeye to send follow-up emails or text messages after a customer's visit, requesting feedback on their experience.
- **Survey Integration:** Include a short, engaging survey link in digital receipts or loyalty program communications to gather insights.
- **Review Monitoring Platforms:** Platforms like Reputation.com or ReviewTrackers can consolidate reviews from Google, Yelp, TripAdvisor, and other sites in one dashboard for easy monitoring.

- **Auto-Responders for Reviews:** Use AI-driven tools to generate personalized responses to online reviews—thanking customers for positive feedback and addressing concerns in a constructive way.
- **Sentiment Analysis:** Leverage analytics tools to analyze trends in customer sentiment, identifying recurring issues or popular aspects of your restaurant.
- **Incentivized Feedback:** Offer discounts or loyalty points for customers who complete feedback surveys or leave a review, encouraging participation.

Making Each Prompt Your Own: Customization Tips

- **Personalize Automated Feedback Requests**: Use tools like Podium or Birdeye to tailor follow-up messages based on customer experiences. For example, after a special event or a new menu launch, send targeted feedback requests asking for opinions on those specific experiences.
- **Engage Customers with Surveys**: Consider integrating short, fun surveys directly into your digital receipts or through your loyalty app, making it easy and rewarding for customers to share their opinions on specific aspects of their visit (e.g., service, food quality, ambiance).
- **Monitor Reviews with a Centralized Dashboard**: Use platforms like Reputation.com or ReviewTrackers to not only track reviews across multiple sites but also analyze customer sentiments. Customize alerts to focus on specific keywords, such as "cocktail" or "service," to keep a pulse on what customers are loving or need improvement.
- **Use AI for Review Responses**: Automate responses to both posi-

tive and negative reviews with an AI-driven tool that customizes the message. Thank loyal customers with personalized follow-up offers or discounts, and address negative reviews with solutions, showing you value constructive feedback.

- **Incentivize Feedback with Rewards**: Motivate customers to leave reviews or complete surveys by offering incentives like a discount on their next visit, free appetizers, or loyalty points. Use ChatGPT to suggest personalized rewards that align with your customer base.
- **Sentiment Analysis for Quick Action**: Use sentiment analysis tools to detect common themes in customer feedback. If customers repeatedly mention slow service or a popular new dish, act quickly by improving service times or promoting the dish more. Automate real-time insights into your restaurant's workflow for better efficiency.
- **Refine Brand Voice in Review Responses**: Customize your auto-responses to reflect your restaurant's tone—whether it's casual and friendly or more formal. If your brand is quirky and fun, let that shine through in your responses to make customers feel even more connected to your restaurant.
- **Requesting More Information:** If you're unfamiliar with any software or app suggestions from ChatGPT, such as 'Podium' or 'ReviewTrackers', simply ask ChatGPT for more details.

By integrating these tailored strategies, you can ensure that customer feedback and review management become seamless, personalized, and efficient, directly contributing to a better customer experience and sustained growth.

Conclusion

In conclusion, with ChatGPT as your financial assistant, managing costs, optimizing profits, and integrating technology becomes significantly more manageable. The prompts in this chapter offer a strong starting point for creating a profitable and sustainable business. By tailoring these strategies to your restaurant's specific needs, you can make well-informed decisions that foster growth and long-term profitability.

8

Menu Descriptions and Updates

In this chapter, we'll explore how to bring fresh and exciting ideas to your restaurant's menu, from unique dishes to seasonal specials. With the help of ChatGPT, you can create dynamic, appealing menu options and craft engaging **Menu Descriptions and Updates** that cater to various customer preferences, dietary needs, and occasions. Whether you're looking for innovative appetizers, beverages, or desserts, we'll show you how to customize prompts to match your restaurant's concept and style, ensuring your menu stays vibrant and relevant. This chapter will empower you to enhance your offerings, keep your guests engaged, and stay ahead in a competitive market.

Crafting Flavorful and Appealing Menu Descriptions

Captivating menu descriptions are essential for enticing guests and showcasing your restaurant's offerings. By using vivid language that highlights ingredients, flavors, and unique aspects of your dishes, you can create an experience that resonates with diners. Customizing descriptions to suit dietary preferences and aligning them with your brand identity not only sparks curiosity but also elevates the overall dining experience.

Prompt 1: Writing a Description for a New Vegetarian Dish

Prompt: *"Write a description for a new vegetarian Thai curry dish featuring tofu, Thai basil, and a spicy green curry sauce."*

Output from ChatGPT

Spicy Green Curry with Tofu – A symphony of bold flavors and vibrant colors, this aromatic green curry combines tender tofu with crisp bell peppers, eggplant, and snow peas, all simmered in a creamy coconut-based green curry sauce infused with fresh Thai basil and fiery green chilies. Served alongside fragrant jasmine rice, it's a plant-based dish that's as comforting as it is adventurous.

Making Each Prompt Your Own: Customization Tips

- **Be Specific:** In your prompt, include the full list of ingredients you want ChatGPT to mention in the dish description. This ensures it only lists the ingredients you specify, rather than making assumptions.
- **Cuisine Style and Atmosphere:** ChatGPT can adjust the tone and language of your dish description to match your restaurant's ambiance. If your restaurant has a casual vibe, ChatGPT can create a more approachable, fun description. For a fine dining experience, ChatGPT can craft an elegant and sophisticated narrative, showcasing the dish's luxurious qualities.
- **Ingredient Sourcing:** If your restaurant prides itself on using locally sourced or organic ingredients, provide this detail to ChatGPT. It can incorporate phrases like "locally grown" or "sustainably sourced" into the dish description, which will not only emphasize quality but also appeal to eco-conscious diners.
- **Dietary Needs:** Let ChatGPT know if your dish caters to specific dietary preferences, such as vegan, gluten-free, or nut-free. ChatGPT can then highlight these attributes in the description, making it clear to guests that your restaurant is attentive to various dietary needs, and broadening the appeal of the dish to a wider audience.
- **Menu Positioning:** Inform ChatGPT whether the dish is a signature item, a seasonal offering, or a special promotion. This allows ChatGPT to phrase the description in a way that gives it a sense of exclusivity and urgency, enticing customers to try it before it's gone or to revisit a fan favorite.

- **Pairings or Meal Context:** If you want to encourage diners to order additional items or elevate the experience, provide ChatGPT with details about recommended pairings. For example, ChatGPT can suggest a wine, beer, or side dish that complements the flavors of your curry, enhancing the overall dining experience and encouraging upsells.

These adjustments help ChatGPT tailor the prompt to your restaurant's specific identity, making the dish description resonate more with your customers.

Prompt 2: Updating the Description for a Classic Pad Thai Dish

Prompt: *"Revamp a menu description for a traditional Pad Thai, making it sound fresh and enticing."*

Output from ChatGPT

Signature Pad Thai – A timeless Thai favorite, this dish features stir-fried rice noodles tossed in a tangy tamarind sauce with bean sprouts, scallions, and crunchy peanuts. Garnished with fresh lime wedges and a touch of chili for a perfect balance of sweet, savory, and spicy. Available with your choice of shrimp, chicken, or tofu.

Making Each Prompt Your Own: Customization Tips

- **Add Proprietary Ingredients or Special Techniques:** If your Pad Thai uses unique ingredients or a special cooking technique, share this with ChatGPT. For example: "Please mention that we use a house-made tamarind sauce to give our Pad Thai a rich, authentic flavor." This helps set your dish apart from others and gives the menu description a unique touch.
- **Customization Options for Protein or Spice Levels:** If your restaurant offers protein choices or allows customers to adjust the spice level, let ChatGPT know so it can incorporate these options. For example: "Make sure to mention that customers can choose between shrimp, chicken, or tofu. Also, highlight that they can customize the spice level to suit their taste, from mild to extra spicy."
- **Cultural or Regional Nuances:** If your Pad Thai recipe is influenced by a specific region or you've added a twist, share that with ChatGPT. For example: "Our Pad Thai is inspired by the coastal flavors of southern Thailand, where seafood is a star ingredient. Add this to the description to make the dish feel more authentic and regionally distinct." This will help emphasize your dish's unique characteristics.
- **Restaurant Theme and Branding:** If your restaurant has a particular theme (e.g., modern fusion, classic Thai street food, fine dining), tell ChatGPT so it can adjust the tone. For example: "Since our restaurant focuses on modern Thai fusion, I want the description to highlight a creative approach to traditional dishes. Make it sound fresh and contemporary, but keep the essence of authenticity." This will ensure the description fits with your brand's vibe.
- **Target Audience Considerations:** Depending on who you're

trying to appeal to, ChatGPT can adjust the language. For example: "We have a lot of health-conscious customers, so please emphasize that this dish is made with fresh, organic ingredients. Also, mention that it's gluten-free and can be made vegan by swapping out the protein." This helps make the dish more appealing to your target demographic.

By clearly telling ChatGPT the desired outcome for each aspect (like ingredients, customization options, or tone), you can craft a menu description that resonates with your restaurant's brand and meets the specific needs of your diners.

Creative Menu Ideas

Creative menu ideas bring excitement and originality to your offerings, helping your restaurant stand out. By experimenting with bold flavor combinations, innovative cooking techniques, or unexpected ingredient pairings, you can create memorable dishes that capture guests' attention and leave a lasting impression.

Prompt 3: Creative Starters Inspired by Thai Street Food

"See Bonus Chapter for an Expanded Version"

Prompt: *"Propose five unique appetizer ideas inspired by Thai street food for a modern restaurant menu."*

Output from ChatGPT

- **Crispy Thai Tacos** – Mini crispy wonton shells filled with spicy minced pork larb, topped with fresh herbs and pickled red onions.
- **Tom Yum Shrimp Skewers** – Juicy shrimp marinated in Tom Yum spices, grilled and served with a zesty lime dipping sauce.
- **Thai Popcorn Chicken** – Bite-sized chicken pieces tossed in a crispy lemongrass batter, served with a sweet chili glaze.
- **Coconut Curry Dumplings** – Steamed dumplings filled with a creamy coconut and vegetable curry, paired with a tamarind-peanut dipping sauce.
- **Street-Style Thai Corn Fritters** – Crispy corn fritters with a hint of red curry, served with a refreshing cucumber relish.

Making Each Prompt Your Own: Customization Tips

- **Serving Style:** Let ChatGPT know if you prefer shareable platters or small plates to encourage communal dining.
- **Spice Levels:** Specify if your audience prefers mild, medium, or bold flavors to ensure the dishes suit their tastes.
- **Presentation:** Specify if you want dishes to have a sleek, modern

look with minimalist plating or bold, artistic presentations. Ask for refined garnishes, unique serving vessels, or interactive elements that enhance the dining experience. For example, microgreens, edible flowers, or creative sauce drizzles can add a sophisticated touch while maintaining the dish's authenticity.

- **Dietary Preferences:** Ask for vegetarian, gluten-free, or other dietary-friendly options to cater to diverse diners.
- **Brand Alignment:** Ensure suggestions match your restaurant's concept, whether it's traditional, fusion, or modern street food.
- **Bonus Chapter:** Have a look at the bonus chapter where you can learn how to expand on this prompt and really make it your own.

By providing ChatGPT with specifics about your restaurant's style, customer preferences, and desired presentation, you can ensure the suggested appetizer ideas align with your vision and brand.

Prompt 4: Brainstorming Unique Thai-Inspired Beverages

Prompt: *"Generate three creative Thai-inspired drink ideas for a summer menu."*

Output from ChatGPT

1. **Lemongrass Mojito** – A refreshing cocktail blending white rum, muddled lemongrass, lime juice, and fresh mint leaves, topped with soda water.
2. **Thai Iced Tea Float** – Traditional Thai iced tea topped with a

scoop of creamy coconut ice cream for a decadent twist.

3. **Spiced Mango Cooler** – Fresh mango puree, lime juice, and soda infused with a touch of bird's eye chili for a subtle kick.

Making Each Prompt Your Own: Customization Tips

- **Specify Alcoholic or Non-Alcoholic Options:** If your menu caters to both drinkers and non-drinkers, tell ChatGPT: "Suggest a mix of cocktails and mocktails, ensuring each has a balanced flavor profile." This helps generate a diverse selection suited to your audience.
- **Highlight Seasonal or Local Ingredients:** Let ChatGPT know if you prioritize fresh, local ingredients. For example: "Incorporate tropical fruits like lychee, passion fruit, and pineapple, which are popular with our customers." This ensures the drinks align with your sourcing strategy.
- **Tailor to Your Restaurant's Theme:** Whether your venue is casual or upscale, guide ChatGPT accordingly. You might say: "Create drinks with an elegant presentation, suitable for a high-end cocktail lounge." This ensures the drinks match your brand identity.
- **Experiment with Unique Presentations:** If presentation is a priority, specify it. For example: "Include creative serving suggestions like smoked glassware, edible flowers, or Thai-inspired garnishes." This will generate visually striking options.
- **Consider Pairing Suggestions:** If you want drinks that complement your menu, tell ChatGPT: "Suggest beverages that pair well with spicy Thai dishes, balancing heat with refreshing flavors." This ensures the drinks enhance the dining experience.

By providing ChatGPT with details about your drink program, you can ensure the beverage ideas are tailored to your restaurant's style and customer preferences.

Seasonal or Special Dish Ideas

Seasonal and special dish ideas offer the perfect opportunity to showcase fresh, locally sourced ingredients while keeping your menu exciting and relevant. Whether it's a dish inspired by a holiday, a limited-time offering, or a celebration of a specific season, these items create urgency and entice guests to try something new.

Prompt 5: Creating a Winter Menu Featuring Thai Comfort Food

Prompt: *"Suggest four hearty Thai dishes perfect for a winter menu."*

Output from ChatGPT

1. **Massaman Beef Curry** – Slow-cooked tender beef simmered in a rich Massaman curry sauce with potatoes, peanuts, and warming spices like cinnamon and cardamom.

2. **Tom Yum Soup** – A spicy and tangy soup with mushrooms, tofu, and lemongrass, served piping hot to chase away the winter chill.
3. **Roasted Duck Red Curry** – Succulent roasted duck breast in a creamy red curry sauce with lychee, pineapple, and Thai basil.
4. **Sticky Rice Pudding** – Warm sticky rice infused with coconut milk and served with caramelized bananas and a sprinkle of sesame seeds.

Making Each Prompt Your Own: Customization Tips

- **Emphasize Hearty, Warming Ingredients:** If you want dishes with rich, warming flavors, tell ChatGPT: "Focus on slow-cooked curries, soups, and stews that feature coconut milk, warming spices, and braised meats." This ensures the menu feels cozy and satisfying.
- **Incorporate Seasonal or Local Produce:** If you source locally, specify: "Use winter vegetables like eggplant, sweet potatoes, and mushrooms in the dishes." ChatGPT can tailor suggestions to what's fresh and available.
- **Adjust for Dining Style (Casual vs. Upscale):** A casual eatery might prefer hearty noodle bowls, while fine dining might need refined plating. Guide ChatGPT with: "Suggest elegant, plated Thai comfort dishes suitable for a fine-dining winter menu."
- **Offer Family-Style or Sharing Options:** If your restaurant encourages communal dining, tell ChatGPT: "Include dishes that work well for family-style sharing, like large-format curries or hot pots." This makes the menu more interactive.
- **Pair with Seasonal Beverages:** To create a well-rounded dining experience, say: "Suggest hot or spiced drink pairings, like Thai tea lattes or mulled cocktails, to complement the winter menu." This

helps boost beverage sales.

By guiding ChatGPT with these details, you can create a Thai winter menu that aligns with your restaurant's unique style and seasonal focus.

Prompt 6: Brainstorming Limited-Time Specials for Valentine's Day

Prompt: *"Generate romantic, Thai-inspired dishes for a Valentine's Day menu."*

Output from ChatGPT

1. **Rose Petal Tom Kha Soup** – A creamy coconut soup with hints of lemongrass and edible rose petals for an elegant twist.
2. **Lover's Duo Platter** – Grilled lobster tails and filet mignon paired with a spicy green curry sauce and sticky rice.
3. **Chocolate Thai Tea Cake** – A rich chocolate cake infused with Thai tea flavors, topped with a coconut cream glaze and gold leaf.

Making Each Prompt Your Own: Customization Tips

- **Specify the Level of Romance and Elegance:** If your restaurant is fine dining, tell ChatGPT: "Suggest upscale, visually stunning Thai dishes with luxurious ingredients like lobster, saffron, and truffle." For a more casual vibe, request: "Create playful, shareable Thai dishes with a fun Valentine's theme."

- **Highlight Aphrodisiac Ingredients:** To enhance the romantic appeal, guide ChatGPT with: "Incorporate ingredients known for their aphrodisiac properties, such as chili, ginger, coconut, dark chocolate, or oysters." This adds a sensual element to the menu.
- **Tailor the Presentation for a Valentine's Feel:** Ask ChatGPT: "Describe unique plating ideas, such as heart-shaped garnishes, edible flowers, or red and pink color themes." This ensures the dishes visually complement the holiday.
- **Choose Between Prix-Fixe or À La Carte Options:** If offering a set menu, specify: "Suggest a cohesive three-course Thai Valentine's Day prix-fixe menu, including an appetizer, entrée, and dessert." If customers can order individually, request: "Provide a mix of indulgent dishes that work well as standalone specials."
- **Incorporate Thai Flavors into Classic Romantic Dishes:** For a creative fusion approach, direct ChatGPT: "Reimagine classic Valentine's dishes like surf and turf or molten chocolate cake using bold Thai flavors and ingredients." This makes the menu both familiar and unique.
- **Consider Beverage Pairings for a Full Experience:** If your restaurant offers cocktails or mocktails, include: "Suggest Thai-inspired drinks that pair well with the dishes, like a lychee and rose spritz or a spiced Thai Old Fashioned." Thoughtful pairings can enhance the meal.

By customizing the prompt with these details, you'll create a Valentine's Day menu that feels personal to your restaurant, wowing guests with a romantic and memorable dining experience.

Innovative Dessert and Snack Ideas

Innovative dessert and snack ideas can elevate your menu by offering unique, crave-worthy options that surprise and delight customers. From unexpected flavor combinations to creative presentations, these items can be both indulgent and refreshing, giving guests an unforgettable finish to their meal or a satisfying snack to enjoy between courses.

Prompt 7: Creating Thai Fusion Desserts

"See Bonus Chapter for an Expanded Version"

Prompt: *"Propose three fusion dessert ideas combining Thai and Western influences."*

Output from ChatGPT

1. **Thai Tea Tiramisu** – Layers of mascarpone cream and Thai tea-soaked ladyfingers, dusted with cocoa and spiced with cardamom.
2. **Mango Cheesecake Bars** – Creamy cheesecake with a tropical mango swirl, set on a buttery coconut cookie crust.
3. **Pandan Crème Puffs** – Light choux pastry filled with a sweet pandan custard and drizzled with white chocolate.

Making Each Prompt Your Own: Customization Tips

- **Align Flavors with Your Existing Menu:** Tell ChatGPT: "Suggest fusion desserts that complement our current Thai dishes, such as incorporating coconut, mango, or Thai tea to match our flavor profiles." This ensures the desserts feel like a natural extension of your menu.
- **Specify the Level of Innovation:** If you want subtle fusion, ask: "Blend Thai flavors into classic Western desserts in a way that feels familiar." For something bolder, request: "Create unique, unexpected Thai fusion desserts that surprise guests."
- **Cater to Dietary Preferences:** Direct ChatGPT with: "Provide gluten-free or dairy-free variations of these desserts so they appeal to a wider audience." You can also request vegan adaptations using coconut cream, cashew-based cheesecake, or dairy-free chocolate.
- **Incorporate Signature Thai Ingredients:** Ask ChatGPT: "Emphasize classic Thai flavors such as pandan, lemongrass, tamarind, or durian, while maintaining an appealing balance for Western palates." This keeps the fusion authentic.
- **Focus on Presentation for a 'Wow' Factor:** To enhance visual appeal, specify: "Describe plating ideas that make the desserts Instagram-worthy, using elements like edible flowers, gold leaf, or coconut ash for contrast."
- **Consider Seasonal or Limited-Time Specials:** If launching these as part of a seasonal menu, guide ChatGPT with: "Suggest Thai fusion desserts that incorporate seasonal fruits or align with holidays like Valentine's Day or summer promotions."
- **Bonus Chapter:** For a more in-depth look at how this prompt can work you and your business don't forget to check out the bonus chapter.

By giving ChatGPT these custom details, you'll create fusion desserts that feel distinct to your restaurant, blending Thai flavors with Western classics in a way that excites customers.

Prompt 8: Developing Thai-Inspired Snacks for a Bar Menu

Prompt: *"Suggest five bar snacks with a Thai twist."*

Output from ChatGPT

1. **Spiced Cashew Mix** – Roasted cashews coated in a blend of lime leaf, chili, and tamarind.
2. **Mini Thai Curry Pies** – Flaky pastry shells filled with a rich green curry chicken mixture.
3. **Tamarind-Glazed Wings** – Crispy chicken wings tossed in a tangy tamarind-chili glaze.
4. **Coconut Shrimp Bites** – Bite-sized shrimp coated in a crispy coconut crust, served with a sweet chili dip.
5. **Lemongrass Fries** – Crispy fries dusted with lemongrass salt and served with a creamy sriracha mayo.

Making Each Prompt Your Own: Customization Tips

- **Align with Your Bar's Drink Offerings:** Tell ChatGPT: "Suggest Thai bar snacks that pair well with our signature cocktails, such as a citrusy lemongrass gin cocktail or a smoky Thai chili margarita." This ensures harmony between food and drinks.
- **Specify Your Bar's Atmosphere:** If your bar is upscale, request: "Create refined, small-plate Thai snacks with elegant plating." For a casual setting, ask: "Suggest fun, hands-on Thai-inspired bar bites that are easy to share."
- **Highlight Local or House-Made Ingredients:** Guide ChatGPT with: "Incorporate locally sourced ingredients or house-made sauces, like our in-house tamarind glaze or Thai chili-infused honey." This personalizes the menu.
- **Cater to Dietary Preferences:** Request options like: "Include at least two vegetarian or vegan Thai-inspired bar snacks that don't compromise on bold flavors." You can also ask for gluten-free options using rice flour or nut-based coatings instead of breadcrumbs.
- **Optimize for Efficient Prep and Service:** If speed is important, specify: "Suggest Thai bar snacks that can be prepped ahead and assembled quickly during service." This helps maintain smooth kitchen operations.
- **Encourage Unique Presentation and Plating:** For a standout menu, ask: "Describe visually appealing plating ideas using banana leaves, Thai ceramic bowls, or mini bamboo baskets to enhance the Thai street food vibe."

By tailoring these details in your ChatGPT prompt, you can craft Thai-inspired bar snacks that align with your bar's identity while offering a unique twist on classic drinking bites.

Cultural and Festival-Specific Menus

Tap into the spirit of tradition by offering cultural and festival-specific menus that celebrate special holidays and unique community celebrations. These menus highlight authentic dishes, seasonal ingredients, and festive flavors, providing your guests with a memorable dining experience that connects them to the richness of diverse cultural heritage and celebrations.

Prompt 9: Designing a Thai New Year (Songkran) Menu

Prompt: *"Create a festive three-course menu inspired by traditional Thai New Year (Songkran) celebrations, focusing on refreshing and celebratory flavors."*

Output from ChatGPT

- **Starter:** Green Mango Salad – Crisp green mango tossed with dried shrimp, roasted peanuts, fresh herbs, and a tangy lime dressing.
- **Main Course:** Grilled River Prawns with Spicy Nam Jim – Juicy grilled river prawns served with a vibrant chili-lime dipping sauce, complemented by sticky rice and grilled vegetables.
- **Dessert:** Coconut Jelly with Seasonal Fruits – Light and refreshing coconut jelly paired with slices of ripe mango, pineapple, and

dragon fruit for a sweet, cooling finish.

Making Each Prompt Your Own: Customization Tips

(The examples below are based on the 'Thai Songkran Festival', but you can easily tailor the information you provide to ChatGPT to suit the festival you're promoting. For instance, if you run a Chinese restaurant, you could adjust it for 'Chinese New Year', or if you have a Mexican restaurant, you could customize it for 'The Day of the Dead' celebrations).

- **Incorporate Traditional Songkran Dishes:** Tell ChatGPT: "Suggest dishes commonly enjoyed during Songkran, such as Khao Chae (jasmine rice in chilled aromatic water) or festive Thai desserts like Thong Yod (golden egg yolk drops)." This adds cultural authenticity.
- **Emphasize Cooling and Refreshing Elements:** Since Songkran is known for water-themed celebrations, request: "Include dishes that are hydrating and cooling, like fruit-based desserts, herbal drinks, or light, citrusy flavors."
- **Align the Menu with Your Restaurant's Concept:** If you run a fine-dining restaurant, ask: "Provide Songkran-inspired dishes with an elegant, plated presentation." For a casual eatery, specify: "Suggest festive, shareable Thai dishes suitable for a lively celebration."
- **Suggest Creative Drink Pairings:** Enhance the experience by requesting: "Include Thai-inspired beverage pairings, such as butterfly pea flower iced tea, lemongrass mojitos, or coconut-infused cocktails."

- **Offer Customization for Dietary Preferences:** If your audience includes vegetarians or health-conscious diners, ask: "Provide plant-based or lighter alternatives for traditional Songkran dishes without compromising flavor."
- **Incorporate Interactive or Shareable Elements:** To make the menu festive, specify: "Suggest dishes that encourage communal dining, such as family-style platters or DIY components like build-your-own lettuce wraps."

Prompt 10: Creating a Thai Vegetarian Festival Menu

Prompt: *"Design a vegetarian menu inspired by Thailand's annual Nine Emperor Gods Festival, focusing on plant-based dishes with bold flavors."*

Output from ChatGPT

- **Starter:** Crispy Tofu and Herb Spring Rolls – Fresh rice paper rolls filled with crispy tofu, Thai basil, mint, and crunchy vegetables, served with a spicy tamarind-peanut dip.
- **Main Course:** Plant-Based Massaman Curry – A rich and aromatic curry featuring hearty chunks of pumpkin, sweet potatoes, and peanuts, served with steamed jasmine rice.
- **Dessert:** Banana and Coconut Milk Soup – Warm coconut milk soup infused with pandan and ginger, served with tender banana slices and a sprinkle of toasted coconut.

Making Each Prompt Your Own: Customization Tips

(The examples below are based on Thailand's annual 'Nine Emperor Gods Festival', but you can easily tailor the information you provide to ChatGPT to suit the festival you're promoting. For instance 'Christmas' or an annual 'Vegan Festival').

- **Focus on Authentic Vegetarian Festival Dishes:** Tell ChatGPT: "Suggest dishes traditionally eaten during the Thai Vegetarian Festival, such as Jay-style (strict vegan) curries, stir-fries, or tofu-based dishes."
- **Highlight Unique Plant-Based Ingredients:** Request: "Incorporate protein-rich plant-based alternatives like jackfruit, tempeh, or banana blossom to replace meat while maintaining bold flavors."
- **Ensure the Menu is Fully Vegan-Friendly:** Since the festival follows strict Buddhist vegetarian guidelines, specify: "Avoid animal products like fish sauce, use alternatives like soy sauce and fermented soybean paste."
- **Match the Menu to Your Restaurant's Theme:** If your restaurant specializes in fusion cuisine, ask: "Provide Thai-inspired vegetarian dishes with global influences, such as a Massaman curry flatbread or Tom Yum risotto."
- **Suggest Creative Plating and Presentation Ideas:** Request: "Include modern plating techniques, such as deconstructed desserts or vibrant color contrasts, to elevate the dining experience."
- **Emphasize Sustainability and Local Ingredients:** If sustainability is part of your restaurant's ethos, guide ChatGPT with: "Suggest vegetarian dishes that use locally sourced, organic produce to minimize environmental impact."

By tailoring your ChatGPT request with these details, you can create a Vegetarian Festival menu that aligns perfectly with your restaurant's identity and customer preferences.

Conclusion

Menu and product creation is a vibrant, ever-evolving process that reflects the heart and soul of your restaurant. With ChatGPT, you can streamline brainstorming, elevate your menu descriptions, and introduce creative new dishes that keep your offerings fresh and appealing.

From seasonal specials to innovative fusion ideas, the prompts in this chapter provide a wealth of inspiration to help you craft a menu that stands out. By tailoring the outputs to your restaurant's unique identity, you can ensure every dish tells a story and every menu item delights your customers.

9

Bonus Chapter

In this **Bonus Chapter**, we'll expand on the prompts from earlier in the book, diving deeper into how you can customize them to better fit your restaurant's specific needs. **This section will show you how small adjustments can make a big impact, ensuring that each prompt works efficiently for your operations.** By tailoring these prompts to your unique business, you'll save time, streamline processes, and enhance communication. Let's take a closer look at how these expanded prompts can work for you!

Prompt 2 from Chapter 4: Marketing and Promotions

TikTok Script for a Behind-the-Scenes Video

Original Prompt: *"Create a TikTok script showcasing how we prepare our signature Pad Thai. Include engaging captions and trending hashtags."*

Example Expanded Prompt

"Create a detailed TikTok script showcasing how we prepare our signature Pad Thai from start to finish. Include engaging and descriptive captions, a clear voiceover that highlights the unique cooking process, and trending hashtags. The video should emphasize the fresh ingredients, the sizzling wok, the cooking techniques, and the final plated dish. Also, suggest a Call to Action (CTA) that encourages viewers to visit the restaurant and mention the video for a discount. Include details on how to incorporate popular TikTok trends, sounds, or challenges to boost engagement."

Example Output from ChatGPT

🎥 **Opening shot**: The camera pans over a bustling kitchen. The sizzling sound of a hot wok fills the air as fresh noodles are tossed in the air by the chef.

🎤 **Voiceover**: "Ever wondered how our signature Pad Thai comes to life? Well, wonder no more! Let's take you behind the scenes and into the heart of our kitchen!"

🎥 **Cut to**: Close-up shot of a chef's hands expertly adding fresh ingredients into the wok—cracked eggs, peanuts, bean sprouts, and a dash of our secret Pad Thai sauce.

🎤 **Voiceover**: "It all starts with fresh ingredients and a perfect balance of flavors. Each ingredient is carefully chosen to bring you the perfect combination of sweet, salty, sour, and spicy. 🍜"

🎥 **Cut to**: The chef stirring the wok with a quick, fluid motion, flames rising up around the noodles as they sizzle.

🎤 **Voiceover**: "The magic happens right here, in the wok. The intense heat locks in all those vibrant flavors, creating the perfect Pad Thai experience!"

🍴 **Caption**: "From wok to plate—our Pad Thai is a symphony of flavors, cooked to perfection. You won't find anything like it anywhere else! Ready to try it for yourself? Come visit us today and taste the magic! 🌟 #BehindTheScenes #PadThaiPerfection"

🎶 **Trending Hashtags**: #BehindTheScenes #ThaiFoodLove #PadThaiPerfection #AuthenticFlavors #FoodieAdventures

🎥 **Final shot**: The finished Pad Thai is plated beautifully, garnished with fresh lime and cilantro, ready to be served to a delighted customer.

🎤 **Voiceover**: "Come and get your hands on this masterpiece, only at [Restaurant Name]. Don't wait—your Pad Thai is waiting for you!"

Making Each Prompt Your Own: Customization Tips

- **Highlight Unique Cooking Techniques:** Provide ChatGPT with details about your restaurant's cooking style and any special techniques you use that make your dishes stand out. For example, you could say, "Our Pad Thai is stir-fried in a traditional wok over high heat, giving it that perfect smoky flavor that you won't find anywhere else." If you have any unique or signature cooking methods (e.g., grilling over an open flame, hand-picking fresh herbs, or using a special blend of spices), share those with ChatGPT to make the message more authentic to your restaurant's cooking philosophy.

- **Add a Unique Call to Action (CTA):** Let ChatGPT know how you'd like to encourage engagement with your audience. For example, "We're offering a special deal for our followers! Come visit us today and mention this video to get 10% off your first Pad Thai order." The more specific you are about the offer or action you want customers to take (whether it's visiting your restaurant, sharing the video, or following your social media), the more customized and effective the CTA will be.
- **Incorporate Trending Challenges or Sounds:** Inform ChatGPT about any current TikTok or social media trends that align with your restaurant's vibe. If there's a trending sound or challenge that you think would work well for your restaurant's theme, tell ChatGPT, like "Use a popular TikTok cooking sound and match it with the motions of making our signature Pad Thai." You could also mention any viral cooking challenges or movements that might fit your restaurant's personality. Sharing such trends with ChatGPT will allow it to suggest ways to weave them into your restaurant's unique content.
- **Showcase Your Brand's Personality:** If your restaurant has a specific atmosphere, make sure ChatGPT knows so the script matches your brand. For example, "We have a relaxed and casual vibe, so ChatGPT should make the voiceover fun and upbeat to match our laid-back atmosphere." If your restaurant is fine-dining, you could request a more polished and elegant script, like "ChatGPT, make sure the tone is refined and sophisticated for our upscale dining experience."
- **Focus on What Makes Your Dish Unique:** ChatGPT can help

spotlight what makes your dish special. For example, "We use fresh, locally sourced ingredients, so ChatGPT should mention that in the script, like 'We use farm-fresh vegetables in our Pad Thai, giving it a taste that's truly local.'" This adds authenticity to the video and highlights the care and quality behind your food.

Keep in mind, the more specific and detailed the information you provide to ChatGPT about what you'd like it to do, the more accurate and helpful the response will be.

Prompt 8 from Chapter 4: Marketing and Promotions

Anniversary Celebration Post

Original Prompt: *"Write a post celebrating the restaurant's 5th anniversary and offering a special discount."*

Example Expanded Prompt

"Write a celebratory social media post for our restaurant's 5th anniversary, offering a special discount to customers. Include the following details:

- *The discount amount is 15% off all meals.*
- *The promotion runs for one week, from [start date] to [end date].*
- *Add a heartfelt thank-you message to customers for their support over the years.*

- *Include a short story or highlight about our restaurant's journey (e.g., how it started, a milestone moment, or customer favorites).*
- *Use a warm, friendly, and celebratory tone.*
- *Encourage customers to visit during the promotion and tag their friends.*
- *Incorporate a unique anniversary hashtag, such as #FiveYearsOfFlavor.*
- *Mention additional celebrations, like a free dessert for customers dining in during the anniversary week.*
- *Add a call-to-action to reserve a table via [website] or call [phone number].*

Make the post visually engaging, and suggest including photos of popular dishes, the restaurant interior, or a collage of memorable moments from the past five years. Be sure to reflect the excitement and gratitude of this milestone!"

Example Output from ChatGPT

🎉 **We're Turning 5!** 🎉

Five incredible years, countless meals, and so many unforgettable memories—all thanks to YOU, our amazing customers! We couldn't have reached this milestone without your love and support.

To celebrate, we're offering **15% OFF** your meal this week from **[start date] to [end date]**. Whether you're a regular or this is your first visit, come join us as we mark this special occasion with great food, even better company, and the same warm hospitality you've come to love. 🥢❤️

It all started five years ago with a dream and a passion for sharing authentic flavors from Thailand with our community. Since then, we've had the privilege of serving you delicious meals, hearing your stories, and becoming a part of your celebrations. From our signature Pad Thai to our much-loved Coconut Sticky Rice with Mango, your support has made all of this possible. 🍜

But wait—there's more! 🎁 In addition to the discount, we're treating all in-house diners to a **FREE dessert** with their meal! Because what's a celebration without a sweet ending? 🍰✨

Make sure to tag your foodie friends and come celebrate with us. We'd love to see your favorite dishes, so don't forget to snap a pic and use our anniversary hashtag: **#FiveYearsOfFlavor.** 📸

Reservations are highly recommended! Book your table today at **[website link]** or call us at **[phone number]** to ensure you don't miss out on the fun.

Here's to many more years of great food, laughter, and memories. Thank you for making the last five years so incredible—we couldn't have done it without you. 💖

Making Each Prompt Your Own: Customization Tips

- **Personalize the Story:** Share specific details about your restaurant's journey with ChatGPT, such as when and how your restaurant started, any milestones you've reached (like opening your first location or winning a local award), and how it's evolved. The more specific you are, the more personal the message will feel. For example, you could provide ChatGPT with information like, "We opened our doors with a small team and a big passion for bringing authentic flavors to the community. Over the years, we've expanded our menu and have been proud to see our loyal customers grow."
- **Highlight Special Dishes or Menu Items:** Let ChatGPT know if your restaurant has any signature dishes that are tied to your

identity. If you have a fan-favorite dish like Pad Thai, a special curry, or a house signature, share these details with ChatGPT. You might say, "Our Spicy Green Curry has been a beloved staple from day one, and it's something we're excited to share with you during our anniversary celebration."

- **Promote Specific Deals:** If your anniversary celebration includes special offers or exclusive deals, tell ChatGPT the specifics. Whether it's a set menu, limited-time specials, or discounts, the more details you provide, the better. For example, you could provide, "We're offering a 15% discount on our anniversary feast, which includes a handpicked selection of our best-selling dishes."
- **Add a Personal Thank You:** To make the message more engaging and personal, tell ChatGPT to express your gratitude to customers directly. You can provide details like, "We are incredibly thankful for our community's support over the years. Without you, we wouldn't be where we are today."
- **Incorporate Your Restaurant's Brand and Tone:** Let ChatGPT know the tone that best represents your restaurant's atmosphere. If your establishment is casual and fun, you can ask for a more lighthearted tone. If your restaurant has a more formal, upscale vibe, ask ChatGPT to craft a more elegant message. Provide specific examples of how your brand communicates, like, "We like to keep things friendly and laid-back, with a touch of humor, but always warm and welcoming."
- **Add Visuals and Hashtags:** Share details with ChatGPT about photos or visuals that would best showcase your restaurant's history. If you've hosted charity events or memorable customer experiences, let ChatGPT know so it can suggest how to incorporate them. Don't forget to mention any unique anniversary hashtags you're using, like #CheersToFiveYears or #[RestaurantName]Anniversary, to encourage engagement on social media.

- **Include Important Details:** To ensure all logistical information is clear, provide ChatGPT with the key event details, such as the dates, location, reservation process, and anything else your customers need to know. For example, let ChatGPT know, "Our celebration will run from [start date] to [end date] with open reservations from [hours]."
- **Give It a Personal Touch:** If you'd like to include any team shout-outs or personalized messages, like a note from the chef or staff, provide ChatGPT with that information. A message such as, "Our entire team has poured their heart into this celebration, and we can't wait to share this special moment with you," would create a connection with your customers.
- **Encourage Social Engagement**: In addition to your hashtag, you could invite customers to share their memories of dining at your restaurant over the past five years. For example: "We'd love to hear your favorite moments with us! Share a memory, a photo, or a dish that you've loved the most using #FiveYearsOfFlavor, and you could win a special gift on your next visit!"

By tailoring these elements to reflect your restaurant's unique story and atmosphere, you'll create a more engaging and memorable anniversary post that will resonate deeply with your customers.

Prompt 4 from Chapter 5: Customer Engagement

Anniversary Congratulations

Original Prompt: *"Write a congratulatory message for a couple celebrating their anniversary at the restaurant."*

Example Expanded Prompt

"Write a congratulatory message for a couple celebrating their anniversary at the restaurant. The couple will be seated in a romantic, private corner near the window with candlelight, flowers, and a beautiful view. Mention that a personalized dessert has been prepared for them, and offer them a complimentary bottle of wine as part of their celebration. The message should express warmth, appreciation, and excitement about their special day."

Example Expanded Output

"🌹✨ Happy Anniversary, [Couple's Names]! ✨🌹

We are so thrilled to celebrate this special day with you at [Restaurant Name]! Your love story is as beautiful as the sunset view from your private corner table by the window. 🥂

To make this evening unforgettable, we've reserved an intimate, candlelit spot just for you, surrounded by fresh flowers and a stunning view that's as lovely as your journey together. 💐

As a sweet gesture, we've prepared a personalized dessert just for the two of you. Enjoy our [Signature Dessert Name], made with love. 🍰 And, to make this moment even more memorable, we'd like to offer you a complimentary bottle of [Wine Name] to toast to your love. 🍷

Here's to many more years of happiness, laughter, and love! ❤ Thank you for choosing us to be part of your celebration."

Making Each Prompt Your Own: Customization Tips

- **Personalized Touch:** Share specific details about the couple that will make the message feel more personal and heartfelt. For instance, you could provide ChatGPT with their names, favorite dishes, or any memorable moments you've had with them at the restaurant. If they're regular customers, mention any inside jokes or shared experiences that will resonate with them, like "We remember when you first tried our Pad Thai on your first date – it's still one of our favorite stories to tell!" This will allow ChatGPT to craft a message that feels truly tailored to the couple.
- **Special Location Details:** Let ChatGPT know about any unique or romantic spots in your restaurant that would be perfect for an anniversary celebration. If you have a quiet garden terrace, a private dining room, or a table near a scenic feature like a fireplace or fountain, provide those details. You could say, "We've reserved our cozy garden terrace for you, where the soft sounds of water and the candlelight will make for the perfect evening," or "Your table by the fireplace will make for an unforgettable anniversary dinner."
- **Customized Menu Items:** If the couple has a history with specific dishes at your restaurant, share those preferences with ChatGPT so it can mention them in the message. For example, "We know that you both love our signature grilled salmon with lemon butter sauce, so we've included it in your anniversary meal," or "You've always enjoyed pairing our red curry with a glass of chilled Sauvignon Blanc, and we've made that perfect pairing for your celebration

tonight." Offering a personalized food experience will make the evening even more special.

- **Exclusive Offerings:** If your restaurant has any anniversary-specific offerings like special set menus, premium wine pairings, or packages designed for couples, provide ChatGPT with these details so they can be included in the message. For example, "We've created a special anniversary set menu just for you, featuring our most popular dishes and a glass of vintage champagne to celebrate this special day." You can also mention any personalized extras, like "We've also included a complimentary dessert for you both to enjoy."
- **Consider the Ambiance:** Share the details of how you plan to create the perfect atmosphere for the couple's celebration. This can include anything from decorations to music. For instance, "We've arranged for a live acoustic guitar performance to serenade you both while you enjoy your meal," or "Your table will be adorned with fresh roses and candles to set a romantic tone." The more details you can provide about the ambiance—such as lighting, music, or floral arrangements—the more personalized the experience will feel.

By feeding ChatGPT with these rich details, you'll ensure the message is perfectly tailored and memorable for the couple's experience at your restaurant.

Prompt 10 from Chapter 5: Customer Engagement

Inviting VIP Customers to an Exclusive Event

Original Prompt: *"Write a personalized invitation for a VIP customer to attend an exclusive tasting event."*

Example Expanded Prompt

"Write a personalized invitation for a VIP customer to attend an exclusive tasting event. The event will feature 7 new dishes curated by Chef [Chef's Name], who will be presenting the menu himself/herself. The event will be limited to 10 loyal customers and will be free of charge as a thank-you for their continued support. Chef [Chef's Name] will also join guests at the table at certain points to chat about the dishes, the inspiration behind them, and all things food and Thailand. The guests are asked to provide honest feedback on the new menu items in return. The event will be held on [date], and the customers must RSVP by [RSVP date] to secure their spot. The tone should be warm, exclusive, and appreciative of the customer's loyalty."

Example Output from ChatGPT

"Dear [Customer's Name],

We have a very special invitation for you! 🌟 As one of our most loyal customers, we'd love to invite you to an exclusive tasting event at [Restaurant Name] on [date].

This will be a private, intimate gathering with only 10 guests, where you'll get to try 7 **new dishes** from our chef [Chef's Name]'s exciting new tasting menu. 🍽 Not only will you be one of the first to sample these dishes, but [Chef's Name] will also be presenting the food personally! At certain points during the evening, our chef will join you at the table to share insights about the dishes, talk all things food, and dive into the inspiration behind each one—straight from Thailand.

This event is **completely free of charge** as a special thank you for your loyalty and support. We only ask for your honest feedback on the new menu items to help us perfect the dining experience.

Space is extremely limited, with only 10 spots available, so please RSVP by [RSVP date] to secure your place. We can't wait to celebrate with you!

Warm regards,
[Your Name]
[Restaurant Name]"

Making Each Prompt Your Own: Customization Tips

- **Be Specific About the Dishes or Chefs Featured:** Provide ChatGPT with details about the dishes in your new tasting menu to build excitement. Share the names and descriptions of a few standout items that will be featured in the event. For example, "We've crafted a new tasting menu, including our fresh crab spring rolls with a spicy tamarind dipping sauce, and a beautifully seared duck breast with lemongrass and coconut cream." If you have a specific chef leading the event, let ChatGPT know so they can highlight their expertise in the message: "Chef [Chef's Name] will be presenting these exquisite dishes personally, sharing insights into the flavors and techniques behind each creation."
- **Highlight the Exclusivity of the Event:** Make the event feel special by emphasizing its exclusivity. Let ChatGPT know that the event is only available to a select group of people, which will

make the invitation feel more luxurious and exclusive. You might say, "As one of our most loyal customers, you're among only 10 people invited to this private tasting event. It's a rare opportunity to experience our new menu before it's officially launched!"

- **Personalize the Tone Based on Your Restaurant's Brand:** Share the general tone or vibe of your restaurant with ChatGPT so it can adjust the message accordingly. If your restaurant has a more casual, friendly atmosphere, you could say, "We're excited to have you join us for this intimate evening! Chef [Chef's Name] is looking forward to showcasing some new favorites, and we can't wait to share them with you." If your restaurant is upscale or formal, ChatGPT can adjust the language to be more refined: "It is our privilege to invite you to an exclusive tasting of our newly curated menu, crafted with the utmost care and culinary expertise."
- **Ensure That Your Event Location Is Described in Detail:** Let ChatGPT know where the event will take place within your restaurant and describe the location to emphasize the intimate and exclusive atmosphere. For example, "The event will be hosted in our private dining room, tucked away in the corner of the restaurant, where only a few guests will be able to enjoy the evening in privacy and comfort." Or, "We've chosen our garden terrace for this event, with soft lighting and candlelit tables, providing the perfect ambiance for a memorable experience."

By sharing these specific details, restaurant owners or managers can provide ChatGPT with everything needed to create a perfectly customized and engaging event invitation that highlights the exclusivity, dishes, and atmosphere of the experience.

Prompt 1 from Chapter 6: Operational Efficiency

Weekly Inventory Checklist for Perishables

Original Prompt: *"Create a weekly inventory checklist for perishables in a Thai restaurant."*

Example Expanded Prompt

"Create a weekly inventory checklist for perishables in a Thai restaurant. Include all relevant fresh ingredients, proteins, herbs, spices, and other perishable items required for preparing authentic Thai dishes. Ensure the list accounts for the most commonly used perishables and can be adjusted based on the specific menu."

Example Output from ChatGPT

Weekly Inventory Checklist – Perishables
Fresh Vegetables:

- Thai basil
- Lemongrass
- Kaffir lime leaves
- Cilantro
- Bean sprouts
- Eggplant
- Green onions

- Bok choy

Fruits:

- Mangoes
- Limes
- Pineapple
- Tamarind pods

Proteins:

- Chicken breasts
- Shrimp (fresh or thawed)
- Tofu (silken and firm varieties)

Dairy and Eggs:

- Eggs
- Fresh coconut milk (if not canned)

Herbs and Spices:

- Fresh ginger
- Garlic
- Galangal
- Thai bird's eye chilies
- Curry pastes (green, red, yellow—store in the refrigerator after opening)

Frozen Perishables:

- Frozen fish fillets (thawed for prep use)
- Frozen prawns

Miscellaneous:

- Rice noodles (store opened packs properly to avoid moisture)
- Sticky rice (keep in airtight packaging after opening)
- Palm sugar (ensure it's stored to maintain freshness)
- Fish sauce (once opened, refrigerate to preserve quality)

Making Each Prompt Your Own: Customization Tips

- **Specify Your Restaurant's Key Dishes:** Share your restaurant's signature dishes with ChatGPT to ensure the inventory checklist reflects the ingredients needed to prepare these dishes. For example, "Please include the ingredients necessary for our signature [Dish Name], such as [ingredient 1], [ingredient 2], and [ingredient 3], which we use regularly. Make sure these ingredients are featured in the checklist, so we can stay well-stocked for our most popular items."
- **Mention Any Locally Sourced Produce:** If your restaurant prides itself on using local ingredients, provide ChatGPT with a list of these items to ensure they're reflected in the checklist. For example, "We source our [ingredient] locally from [farmer's market/farm name], so be sure to include it in the list for its freshness and to maintain the quality of our dishes."
- **Add Details About Seasonal Dishes or Rotating Menu Items:** If your restaurant offers seasonal or rotating menu items, let ChatGPT know when these items are available so it can adjust

the checklist accordingly. For example, "We have a seasonal dish, [Dish Name], that we serve in the fall and winter, so make sure to include ingredients like [ingredient 1] and [ingredient 2] during those months."

- **Group Items by Storage Requirements:** To streamline inventory management, provide ChatGPT with the preferred storage methods for different ingredients, so it can group them by storage needs. For example, "Please group all perishables, such as fresh produce and dairy, under the 'refrigeration' section, and frozen items like [ingredient] and [ingredient] under 'freezer storage.' This will help us stay organized and reduce spoilage."
- **Incorporate Inventory Quantities or Par Levels:** If you need a more specific inventory system to help with reordering, ask ChatGPT to include inventory quantities or par levels. For example, "Include the current stock levels for each ingredient, and set minimum par levels for items like [ingredient 1] and [ingredient 2], so we know when it's time to reorder."
- **Include Specialty Ingredients or Homemade Items:** If your restaurant uses unique or homemade ingredients, be sure to mention these to make sure they're included in the inventory checklist. For example, "Please add our homemade [sauce name], [broth name], or [marinade name] to the checklist, and note that we prepare them fresh each week, so we'll need to keep track of ingredients like [ingredient] and [ingredient]."

By sharing these specific details with ChatGPT, restaurant owners or managers can receive a customized and comprehensive inventory checklist tailored to their menu, sourcing practices, and operational needs. This will help ensure they have the necessary ingredients stocked and ready for each service.

Prompt 5 from Chapter 6: Operational Efficiency

Weekly Staff Schedule for Kitchen and Front-of-House

Original Prompt: *"Create a weekly schedule for kitchen and front-of-house staff."*

Example Expanded Prompt

"Create a detailed weekly staff schedule for both kitchen and front-of-house, ensuring optimal coverage for all service times, peak hours, and any special events or promotions. The schedule should account for key roles such as chefs, line cooks, servers, hosts, and bartenders, and should consider the restaurant's unique flow, including prep times, busy dinner rushes, and slower hours."

Example Output from ChatGPT

Weekly Staff Schedule – [Restaurant Name]

Monday – Thursday:

Kitchen:

- **Chef [Name]** – 10 AM – 5 PM (Prep & Order Management)
- **Sous Chef [Name]** – 11 AM – 6 PM (Support Chef, Oversee Line)
- **Line Cook [Name]** – 12 PM – 9 PM (Lunch & Dinner Service)
- **Prep Cook [Name]** – 9 AM – 4 PM (Prep Work, Sauces, Garnishes)
- **Dishwasher [Name]** – 12 PM – 8 PM (Maintain Cleanliness)

Front-of-House:

- **Host [Name]** – 11 AM – 3 PM (Lunch Service)
- **Server [Name]** – 12 PM – 9 PM (Lunch & Dinner Service)
- **Bartender [Name]** – 4 PM – 10 PM (Dinner Service)
- **Busser [Name]** – 5 PM – 10 PM (Busy Dinner Rush)

Friday – Saturday (Peak Weekend Hours):
Kitchen:

- **Chef [Name]** – 11 AM – 8 PM (Manage Service, Special Orders)
- **Sous Chef [Name]** – 12 PM – 9 PM (Prepare and Oversee Service)
- **Line Cook [Name]** – 1 PM – 10 PM (Dinner Rush)
- **Prep Cook [Name]** – 10 AM – 5 PM (Ensure Fresh Ingredients for Rush)
- **Dishwasher [Name]** – 2 PM – 11 PM (Constantly Clean & Reset Stations)

Front-of-House:

- **Host [Name]** – 3 PM – 9 PM (Dinner Rush & Table Management)
- **Server [Name]** – 1 PM – 10 PM (Dinner Service, Peak Hours)
- **Bartender [Name]** – 4 PM – 11 PM (Dinner & Bar Service)
- **Busser [Name]** – 5 PM – 11 PM (Assist Servers, Reset Tables)
- **Server [Name]** – 5 PM – 10 PM (Additional Support for Dinner Rush)

Sunday (Brunch + Evening Service):
Kitchen:

- **Chef [Name]** – 9 AM – 5 PM (Brunch Prep, Evening Service)

- **Sous Chef [Name]** – 10 AM – 6 PM (Prepare for Sunday Night Dinner)
- **Line Cook [Name]** – 9 AM – 4 PM (Brunch Rush)
- **Prep Cook [Name]** – 10 AM – 3 PM (Prepare Dishes for Dinner Service)
- **Dishwasher [Name]** – 12 PM – 8 PM (Maintain Cleanliness for Both Services)

Front-of-House:

- **Host [Name]** – 9 AM – 3 PM (Manage Brunch Seating)
- **Server [Name]** – 9 AM – 4 PM (Brunch Service)
- **Bartender [Name]** – 11 AM – 6 PM (Bar & Specialty Drinks for Brunch)
- **Busser [Name]** – 9 AM – 3 PM (Assist during Brunch Rush)
- **Server [Name]** – 3 PM – 9 PM (Evening Service)
- **Bartender [Name]** – 4 PM – 9 PM (Support Evening Drink Service)

Making Each Prompt Your Own: Customization Tips

- **Consider Your Restaurant's Peak Hours:** Share detailed information about your restaurant's busiest times to adjust shift lengths accordingly. For example, "Our dinner rush from 6 PM to 9 PM is our peak period, so I need additional servers and kitchen staff to handle the increased volume. On weekends, our brunch from 10 AM to 2 PM is particularly busy, so please ensure there are enough hosts and servers in place during those hours."
- **Account for Special Events and Promotions:** Notify ChatGPT

of any special events or promotions that might require extra staff. For example, "We're hosting a special holiday menu from [date] to [date], and we'll need additional line cooks, servers, and bartenders. If there's a VIP night on [date], please schedule extra support staff to help manage the crowd."

- **Include Prep Time for Complex Dishes:** If your restaurant offers complex dishes that require longer preparation, let ChatGPT know about the necessary prep times. For example, "Our [Dish Name] requires significant prep time due to its fresh ingredients and specialized preparation techniques. Please schedule a prep cook to start at [time] to ensure everything is ready for service."
- **Rotate Shifts for Employee Work-Life Balance:** For a healthier work-life balance, provide ChatGPT with your preferences for rotating shifts across your team. For example, "Please ensure shifts are rotated fairly so no one is consistently scheduled for late nights or weekends, unless it is requested or agreed upon. I want everyone to have a chance for time off during weekends and holidays."
- **Consider Breaks and Staff Well-being:** Ensure that your staff has proper breaks, especially during long shifts, to maintain morale and performance. For example, "Please build breaks into the schedule for servers after 3 hours of service, and ensure that kitchen staff have at least a 30-minute break during shifts longer than 6 hours."
- **Track Staff Availability and Preferences:** Input your staff members' preferred workdays, times, and any time-off requests, and share this with ChatGPT so the schedule reflects their availability. For example, "Our bartender prefers evening shifts, so please schedule them accordingly. Additionally, one of our servers requested Sundays off, so please ensure they're not scheduled."
- **Include Flexibility for Unexpected Changes:** Since the restaurant industry often deals with unexpected changes, be sure to

ask ChatGPT to include flexibility in the schedule. For example, "Please make sure to have backup staff ready in case of any call-outs or no-shows. It would be great if you could suggest a few flexible shift options for those last-minute changes."

- **Balance Kitchen and Front-of-House Staffing:** Ensure your kitchen and front-of-house staff are balanced to handle peak times. For example, "If we're expecting a large dinner rush, I'll need 3 line cooks and 2 prep cooks on the kitchen side. On the front-of-house, make sure we have enough servers, bartenders, and bussers during peak hours to keep everything running smoothly."

By providing these specific details to ChatGPT, you can ensure that the weekly staff schedule is customized to your restaurant's operational needs, from peak hours to special events, staff preferences, and flexibility for last-minute changes.

Prompt 8 from Chapter 6: Operational Efficiency

Staff Quiz on Food Safety Best Practices

Original Prompt: *"Generate a staff quiz on food safety best practices for kitchen staff."*

Example Expanded Prompt

"Create a staff quiz on food safety best practices for kitchen staff. The quiz should include 6 questions in total and must feature the following four specific questions:

1. *What should you do before handling food?*
2. *What is the proper internal temperature for cooked poultry?*
3. *How should raw meat be stored in the refrigerator?*
4. *What is the correct way to cool down hot food for storage?*

Each question should be multiple choice with three options (labeled a, b, and c). Include one trick question to ensure staff pay close attention to details, and tailor all questions to the needs of a Thai restaurant, focusing on real-life kitchen scenarios."

Example Output from ChatGPT

Food Safety Quiz for Kitchen Staff in a Thai Restaurant

1. What should you do before handling food?

a. Wipe your hands with a towel.

b. Wash your hands thoroughly with soap and water for at least 20 seconds.

c. Use hand sanitizer only.

Answer: b

2. What is the proper internal temperature for cooked poultry, such as chicken used in Pad Thai or Green Curry?

a. 145°F (63°C)
b. 165°F (74°C)
c. 180°F (82°C)
Answer: b

3. How should raw meat, such as beef or chicken, be stored in the refrigerator?
a. On the top shelf, away from the door.
b. On the bottom shelf, in a sealed container, to prevent cross-contamination.
c. Anywhere in the fridge, as long as it's wrapped tightly in plastic.
Answer: b

4. What is the correct way to cool down hot food, such as Tom Yum soup, for storage?
a. Leave it at room temperature for several hours, then refrigerate.
b. Transfer it to smaller, shallow containers and place it in the refrigerator immediately.
c. Put it in a large container, cover it tightly, and place it directly in the freezer.
Answer: b

5. Which of the following is true about cross-contamination? ***(Trick Question)***
a. It only happens when raw meat touches cooked food.
b. Washing cutting boards and knives with hot, soapy water between uses can prevent it.
c. It's not a concern if food is cooked to a high enough temperature.
Answer: b

6. How long can cooked food, like leftover Pad Thai or Massaman

Curry, be safely kept at room temperature?

a. Up to 4 hours.
b. Up to 2 hours.
c. Up to 6 hours.
Answer: b

This quiz ensures kitchen staff are familiar with essential food safety practices tailored to a Thai restaurant's needs, such as handling specific ingredients and working efficiently to maintain food quality and safety.

Making Each Prompt Your Own: Customization Tips

- **Include Specific Ingredients and Scenarios:** Provide ChatGPT with details about the specific ingredients or foods that require special handling in your restaurant. For example, "Include questions about how to handle fresh herbs like Thai basil, lemongrass, and cilantro to prevent spoilage. Ask about the proper storage of seafood, particularly shrimp and fish, to ensure they are stored at the correct temperature and prevent contamination."
- **Add Real-Life Scenarios:** Integrate real-life situations that your kitchen staff is likely to encounter. For example, "Ask questions on how to properly store curry paste or ensure the correct handling of opened coconut milk. For example, 'What's the correct procedure for storing curry paste once opened?' and 'How should opened coconut milk be stored to prevent contamination?'"
- **Emphasize Key Risks:** Focus on the specific risks associated with the menu. For example, "Incorporate questions that emphasize cross-contamination, particularly with raw proteins such as chicken or seafood. Include a scenario where staff needs to ensure rice is properly stored to avoid bacterial growth, with a

question like, 'What is the best method to store cooked rice to prevent bacterial contamination?'"

- **Adjust Difficulty Levels:** Tailor questions for different experience levels. For example, "For more experienced chefs, include questions about handling and preserving complex ingredients, such as Thai curry paste or fish sauce, while for newer staff, focus on basic food safety practices like hand-washing or temperature control."
- **Follow Up with Practical Training:** Ensure that the quiz results inform your next training steps. For example, "Once you assess the quiz results, use the answers to identify specific knowledge gaps and organize follow-up training sessions. If the team struggles with storing raw seafood, you can provide additional training on seafood safety."

By customizing the quiz in this way, you'll ensure it directly addresses your restaurant's unique needs and challenges. This will help you train your team more effectively and enhance food safety standards across the board.

Prompt 1 from Chapter 7: Financial and Business Strategy

Pricing a new menu item based on the cost of ingredients

Original Prompt: *"Analyze the cost of ingredients for a new menu item (Grilled Thai Pork Skewers) and suggest a reasonable price point."*

Example Expanded Prompt

"Analyze the cost of ingredients for a new menu item (Grilled Thai Pork Skewers) and suggest a reasonable price point. Include a detailed breakdown of the ingredient costs, calculate the price using a food cost percentage, and consider adjustments based on the restaurant type (casual dining, fine dining, etc.), market trends, and portion size."

Example Output from ChatGPT

To calculate the cost and suggest a price for your new menu item, let's break it into the following components:

Ingredient Cost Breakdown

- **Pork (per 3 skewers):** $2.50
- **Marinade (fish sauce, garlic, brown sugar, lime juice, herbs):** $0.80

- **Rice (per serving):** $0.50
- **Garnishes (cilantro, peanuts):** $0.30
- **Wastage (e.g., trimming pork or herb stems):** $0.20
- **Total Food Cost (per skewer):** $4.30

Pricing Calculation

Using a standard food cost percentage of 30%, we calculate the price as:

- **Suggested Menu Price = Total Food Cost ÷ Desired Food Cost Percentage**
- Suggested Menu Price = $4.30 ÷ 0.30 = $14.33 (rounded to $14.95).

Market and Competitor Adjustments

- Research shows comparable appetizers in upscale Thai restaurants are priced between $12 and $18.
- To align with the target market, a price of $14.95 is competitive while ensuring profitability.

Portion Size and Value Perception

- Serve 3 skewers per portion, accompanied by a side of jasmine rice and a tamarind dipping sauce.
- For premium presentation, place the skewers on a banana leaf and garnish with lime wedges and crushed peanuts.

Profit Margin Considerations

- Adjust portions or garnishes as needed to achieve a balanced food cost percentage.

- Balance potential lower margins on this dish by upselling accompaniments, like signature cocktails or dessert pairings.

Making Each Prompt Your Own: Customization Tips

Here are ways to provide additional details to ChatGPT for a tailored analysis:

- **Provide Exact Ingredient Prices:** When working with ChatGPT to price your menu item, share the precise cost of each ingredient. For example, "The pork for the skewers costs $2.50, the marinade ingredients total $0.80, and garnishes like cilantro and peanuts add $0.30." This allows ChatGPT to generate a more accurate cost breakdown and price calculation.
- **Highlight Unique Ingredients or Preparation Techniques:** If you're using premium ingredients (like organic pork, locally sourced herbs, or specialty spices), share those details with ChatGPT. For instance, "We use organic pork, which costs slightly more than conventional cuts, and the skewers are marinated for 24 hours." ChatGPT can then incorporate these nuances into the pricing analysis and adjust costs accordingly.
- **Specify Your Restaurant Type and Target Audience:** Mention whether your restaurant is fine dining, casual, or fast-casual. For example, "Our restaurant is casual dining, so we aim for a food cost percentage of around 30%." ChatGPT can tailor the pricing strategy based on your restaurant type and the market you're targeting, adjusting food cost percentages and pricing accordingly.
- **Consider Regional Market and Competition:** Let ChatGPT

know about your location and competitors. For example, "Our competitors in the area charge between $12 and $18 for similar appetizer dishes." This will allow ChatGPT to adjust its pricing suggestion based on the local competitive landscape and customer expectations.

- **Adjust for Portion Size Variations:** Be clear about how you want the dish to be served. If you plan to offer the grilled pork skewers as part of a larger meal or as a shareable appetizer, share that with ChatGPT. For example, "We plan to serve three skewers per portion with a side of jasmine rice." ChatGPT can then recommend pricing for different portion sizes and even suggest upsell opportunities like extra skewers or side sauces.
- **Account for Seasonal Ingredient Cost Fluctuations:** If certain ingredients fluctuate in price seasonally, such as lime or cilantro, provide that information to ChatGPT. For example, "Limes are more expensive during the off-season, so let ChatGPT recommend ways to adjust pricing or portions accordingly during those months." This allows for flexibility in your pricing strategy throughout the year.
- **Reflect Your Restaurant's Presentation and Branding:** Inform ChatGPT about your restaurant's visual style and presentation approach. For example, "We serve our skewers on banana leaves with lime wedges and crushed peanuts for a rustic yet elevated feel." ChatGPT can integrate these visual elements into the pricing analysis, ensuring that the dish is priced appropriately for the style of presentation you envision.
- **Balance Menu Costs and Profitability:** If this dish is a high-cost item, you may need to balance its pricing with higher-margin items. Let ChatGPT know if you're aiming to create a profitable overall menu strategy. For example, "We want to offer affordable appetizers but will pair this dish with signature cocktails to increase

overall profitability." ChatGPT can then suggest pricing strategies that complement the rest of the menu and drive sales.

By providing these details to ChatGPT, you ensure the pricing analysis is highly customized to your restaurant's specific needs, making it easier to arrive at a price point that aligns with your goals.

Prompt 4 from Chapter 7: Financial and Business Strategy

Suggestions for increasing profit margins

Original Prompt: *"Generate suggestions for increasing profit margins without compromising quality."*

Example Expanded Prompt

"Generate suggestions for increasing profit margins without compromising quality. Provide detailed strategies for menu optimization, cost control, labor efficiency, and the use of technology. Tailor suggestions based on the restaurant type (e.g., fine dining, fast-casual, or family-friendly) and specific cuisine focus. Include ways to balance profitability with maintaining customer satisfaction and brand integrity."

Example Output from ChatGPT

Here are detailed strategies for increasing profit margins without compromising on quality:

1. Menu Engineering

- **Optimize High-Profit Items:** Highlight high-margin dishes on your menu by placing them in prominent positions or using eye-catching descriptions. For example, feature your best-selling appetizers in a "Chef's Favorites" section.
- **Remove Low-Margin Items:** Regularly review your menu performance. If certain dishes have low sales or poor margins, either remove them or rework the recipe to make them more cost-effective.
- **Encourage Add-Ons:** Introduce premium upgrades (e.g., truffle oil on fries, extra protein options) or bundling strategies like prix-fixe menus to increase average order values.

2. Improve Portion Control

- **Standardize Portions:** Train kitchen staff to measure portions consistently to minimize waste and control food costs. Use tools like portioning spoons or digital scales.
- **Right-Sizing Plates:** For items where over-portioning is common, consider slightly smaller serving sizes while maintaining strong presentation. This reduces food waste while keeping perceived value high.

3. Labor Efficiency

- **Optimize Scheduling:** Use historical data to identify peak and slow hours. Schedule staff based on demand, avoiding overstaffing during low-traffic periods.
- **Cross-Train Employees:** Ensure staff can perform multiple roles (e.g., servers who can bus tables or bartenders who can manage the register). This flexibility reduces the need for additional hires during peak times.

4. Use Technology for Efficiency

- **Leverage a POS System:** Use POS analytics to track inventory, food waste, and slow-moving items. Adjust purchasing habits based on these insights to reduce overstocking.
- **Online Ordering and Delivery:** Optimize online platforms for delivery and takeout. Introduce delivery-only menu items that have high margins and low prep complexity.
- **Inventory Management Tools:** Automate inventory tracking to identify trends in waste or theft, ensuring your purchasing aligns with real-time demand.

5. Reduce Food Waste

- **Repurpose Ingredients:** Use trimmings or leftovers creatively—for example, repurpose vegetable scraps into broths or sauces.
- **Seasonal Menus:** Incorporate in-season ingredients, which are typically more affordable and fresher. Seasonal specials can also create excitement and drive sales.

6. Enhance Customer Value Perception

- **Focus on Presentation:** Invest in plating techniques that elevate the visual appeal of dishes, making smaller portions appear more luxurious.

- **Create Experiences:** Offer value-added experiences, such as chef's tasting menus or wine pairings, to justify higher prices without altering the portion size.

Making Each Prompt Your Own: Customization Tips

- **Provide Specific Cuisine Focus:** When asking ChatGPT for suggestions, be specific about your cuisine focus. For example, "We're a Thai restaurant, and we want to focus on high-margin street food-inspired appetizers like skewers or fried spring rolls." This helps ChatGPT tailor its recommendations to dishes that suit your menu and cuisine style, ensuring strategies are relevant and cost-effective.
- **Tailor Strategies Based on Restaurant Type:** Let ChatGPT know if your restaurant is fine dining, fast-casual, or family-friendly. For example, "Our restaurant is fine dining, so we want to focus on premium experiences like wine pairings and multi-course meals to justify higher prices." This enables ChatGPT to provide suggestions that align with your specific restaurant type, ensuring the strategies match your service style.
- **Adjust for Regional Market Differences:** Share details about your location and local competition. For instance, "We're located in an urban area, where high-margin specialty cocktails are popular, but in suburban areas, family-style meals are preferred." ChatGPT can then adjust its suggestions to help you stay competitive in your market by considering customer preferences and local pricing norms.

- **Integrate Brand Identity:** If your restaurant has a unique brand or value proposition, let ChatGPT know. For example, "Our restaurant focuses on sustainability and locally sourced ingredients, and we want to emphasize these aspects in our menu pricing and offerings." ChatGPT can then ensure any profit-increasing suggestions align with your brand's message and reputation.
- **Incorporate Seasonal and Signature Items:** If you plan to create seasonal specials or highlight signature dishes, mention it. For example, "We want to feature seasonal specials that use fresh, affordable ingredients like spring vegetables or seafood." ChatGPT can then help you craft a pricing strategy for these items, considering cost fluctuations and maximizing profit during peak seasons.
- **Share Customer Habits and Preferences:** Provide insights into customer habits to help ChatGPT tailor its suggestions. For example, "Our customers frequently order cocktails, and there's a high demand for vegetarian options." ChatGPT can then suggest ways to leverage this customer behavior, such as offering high-margin cocktails or plant-based dishes to increase profit margins.
- **Adjust Portion Sizes for Profitability:** Be specific about portion sizes and how they can impact margins. For instance, "We want to slightly reduce portion sizes for our main dishes to lower food costs but maintain the perceived value." ChatGPT can then recommend portion control strategies that minimize waste while maintaining customer satisfaction.
- **Use Technology to Streamline Operations:** Let ChatGPT know if you're using technology, like POS systems or inventory management tools. For example, "We use a POS system to track inventory and identify slow-moving items." ChatGPT can then suggest ways to use technology for cost control, such as adjusting purchasing habits or optimizing menu offerings based on sales

data.

- **Offer Premium Add-Ons or Upsells:** Mention if you plan to incorporate upselling strategies like premium add-ons. For example, "We want to offer premium options like truffle oil on fries or extra protein for an additional charge." ChatGPT can then recommend upsell tactics that increase average ticket size while enhancing customer experience.
- **Create Experiences to Justify Price Increases:** If you want to justify higher prices through experiences, share that with ChatGPT. For example, "We plan to introduce a chef's tasting menu or wine pairings to offer a more elevated dining experience." ChatGPT can then help you design experiences that enhance the perceived value of your dishes, justifying a higher price point.

By customizing your request with specific details about your restaurant, customer base, and goals, ChatGPT can provide tailored suggestions that directly impact your restaurant's profitability without compromising quality.

Prompt 3 from Chapter 8: Menu Descriptions and Updates

Suggesting Creative Starters Inspired by Thai Street Food

Original Prompt: *"Propose five unique appetizer ideas inspired by Thai street food for a modern restaurant menu."*

Example Expanded Prompt

"Propose five creative and unique appetizer ideas inspired by Thai street food, tailored for a modern restaurant menu. Provide a detailed description for each appetizer, considering the following elements:

1. ***Flavors and Ingredients:*** *Incorporate authentic Thai flavors and ingredients, such as lemongrass, galangal, kaffir lime leaves, tamarind, coconut milk, chili, and fresh herbs. Explain how these ingredients are used to create bold, harmonious flavor profiles.*
2. ***Street Food Inspiration:*** *Draw direct inspiration from traditional Thai street food dishes (e.g., larb, satay, pad kra pao, or khanom jeen). Highlight how the essence of street food is preserved while modernizing the dish for a restaurant setting.*
3. ***Presentation and Plating:*** *Describe visually appealing ways to plate each dish, ensuring the appetizers are Instagram-worthy and aligned with the restaurant's branding. Consider using unique plating styles or creative garnishes to elevate the street-food vibe.*
4. ***Dietary Adaptations:*** *Offer suggestions for making each appetizer adaptable for dietary restrictions, such as gluten-free, vegetarian, or*

vegan options, without compromising authenticity or flavor.

5. ***Shareability and Appeal:*** *Ensure the appetizers are designed to be shareable, easy to eat, and ideal for group dining, which aligns with the communal spirit of Thai cuisine.*
6. ***Practicality for a Restaurant Setting:*** *Provide insights into preparation and service, ensuring that each appetizer is feasible for a busy restaurant kitchen and easy to execute during peak hours.*
7. ***Creative Names and Storytelling:*** *Suggest enticing and memorable names for each appetizer, accompanied by a brief story or cultural reference that connects the dish to Thai street food traditions."*

Example Output from ChatGPT

Spicy Larb Crostini

- **Flavors and Ingredients:** Traditional larb (minced chicken or pork) with a spicy lime dressing, toasted rice powder, and fresh herbs, served on bite-sized crostini for a crisp, modern twist.
- **Street Food Inspiration:** Larb is a staple of Thai street food, transformed into an elegant appetizer here.
- **Presentation and Plating:** Topped with pickled shallots and microgreens, served on a wooden board with lime wedges.
- **Dietary Adaptations:** Substitute tofu for the protein to create a vegan-friendly option.
- **Practicality:** Larb can be prepped ahead and assembled quickly on crostini during service.
- **Name:** *Larb Bites*—a playful name that highlights the dish's street food roots.

Tom Kha Arancini

- **Flavors and Ingredients:** Creamy rice infused with the flavors of Tom Kha (coconut soup with lemongrass and galangal), breaded and fried into crispy arancini balls, served with a chili-lime aioli.
- **Street Food Inspiration:** Inspired by the comforting flavors of Tom Kha soup, reimagined in an Italian fusion format.
- **Presentation and Plating:** Served on a minimalist plate with a drizzle of aioli and a sprinkle of crushed peanuts.
- **Dietary Adaptations:** Use gluten-free breadcrumbs for a gluten-free option.
- **Practicality:** Arancini can be prepped ahead of time and fried to order.
- **Name:** *Coconut Arancini Delight.*

Pad Kra Pao Sliders

- **Flavors and Ingredients:** Mini slider buns filled with stir-fried spicy basil chicken (pad kra pao), topped with a fried quail egg and a drizzle of sweet chili mayo.
- **Street Food Inspiration:** Pad Kra Pao is a beloved street dish in Thailand, made more portable as a slider.
- **Presentation and Plating:** Served on a slate platter with sesame seeds sprinkled on top of the buns and a side of Thai-style slaw.
- **Dietary Adaptations:** Substitute the chicken with jackfruit or plant-based protein for a vegan version.
- **Practicality:** Fillings can be batch-cooked and sliders assembled just before serving.
- **Name:** *Basil Bliss Sliders.*

Coconut and Kaffir Lime Shrimp Cakes

- **Flavors and Ingredients:** Shrimp cakes infused with kaffir lime, coconut, and red curry, served with a tangy tamarind dipping sauce.
- **Street Food Inspiration:** Based on the classic Thai tod mun goong, a popular fried street snack.
- **Presentation and Plating:** Served on a banana leaf-lined plate with dipping sauce in small ceramic bowls.
- **Dietary Adaptations:** Replace shrimp with a blend of chickpeas and sweet potatoes for a vegetarian option.
- **Practicality:** Cakes can be formed ahead of time and fried to order.
- **Name:** *Lime-Coconut Shrimp Cakes.*

Red Curry Roti Rolls

- **Flavors and Ingredients:** Crispy roti bread rolled and stuffed with red curry chicken, crunchy vegetables, and a hint of coconut cream, sliced into bite-sized pieces.
- **Street Food Inspiration:** Combines Thai-style roti and curry into a portable, sharable appetizer.
- **Presentation and Plating:** Arranged in a spiral pattern on a platter with dipping sauce in the center, garnished with Thai basil.
- **Dietary Adaptations:** Use tofu and vegetables for a vegan-friendly version.
- **Practicality:** Can be pre-prepped and reheated without losing quality.
- **Name:** *Curry Roti Rolls.*

Making Each Prompt Your Own: Customization Tips

- **Specify Protein Preferences:** If you want to highlight specific proteins like chicken, shrimp, or tofu, let ChatGPT know. For example, "We want to focus on shrimp-based appetizers for the menu," and ChatGPT can tailor suggestions based on your protein preferences, such as creating a shrimp-based version of a classic Thai dish.
- **Dietary Focus:** Share if your restaurant caters to specific dietary needs. For instance, "We want more vegetarian and vegan options on our menu." ChatGPT can then suggest ways to adapt classic Thai street food dishes to cater to vegan or vegetarian guests, ensuring they're both authentic and flavorful.
- **Plating Style Preferences:** If you'd like a specific plating style, such as traditional Thai serveware or modern minimalist plating, share that with ChatGPT. For example, "We want to use bamboo baskets or banana leaves for plating to add an authentic touch." ChatGPT can incorporate these details into the visual presentation of each appetizer.
- **Restaurant Style and Ambiance:** Indicate whether your restaurant is fine dining, casual, or a different style to align the suggestions with your concept. For example, "Our restaurant has a modern casual vibe, and we want the appetizers to reflect that while still maintaining authentic flavors." ChatGPT can adjust the tone of the dishes and their presentation to fit your restaurant's overall atmosphere.
- **Prep Requirements:** Be specific about whether you want appetizers that can be prepped in advance or made to order. For example, "We have a small kitchen, so we need appetizers that can be prepped ahead of time." ChatGPT can then suggest appetizers that can be easily prepped in bulk and assembled quickly during

service, making them more efficient for busy kitchens.

- **Special Promotions or Events:** If you're planning a special event or promotion, share this with ChatGPT so it can suggest relevant appetizers. For instance, "We're hosting a Thai street food night next month." ChatGPT can recommend appetizers that align with the theme, helping to create a cohesive and appealing menu for the event.
- **Highlight Cultural References:** If you want to incorporate specific cultural elements or stories behind the dishes, mention them in your prompt. For example, "We want each dish to have a story that connects it to Thai street food traditions." ChatGPT can then provide names and backstories for each appetizer, making the menu more engaging and memorable.
- **Customer Preferences or Trends:** If you've noticed certain customer preferences or trends at your restaurant, share them with ChatGPT. For instance, "Our customers love spicy foods, and we're looking for ways to incorporate more heat into the appetizers." ChatGPT can suggest appetizers with bold flavors or more spicy elements to cater to customer tastes.

By providing these specific details to ChatGPT, you can ensure that the suggested appetizers are tailored to your restaurant's needs, from flavors to plating, while maintaining the authenticity of Thai street food.

Prompt 7 from Chapter 8: Menu and Product Creation

Creating Thai Fusion Desserts

Original Prompt: *"Propose three fusion dessert ideas combining Thai and Western influences."*

Example Expanded Prompt

"Develop three creative fusion dessert ideas combining Thai and Western influences, with detailed descriptions for each dish. Include the following details:

1. ***Flavors and Ingredients:*** *Highlight the key Thai and Western flavor profiles being blended. Incorporate traditional Thai ingredients (e.g., pandan, coconut, lemongrass, Thai tea, mango, tamarind) and explain how they pair with classic Western dessert components.*
2. **Presentation and Plating:** *Describe how the dessert would be presented, including plating styles or unique garnishes, ensuring the dish is visually striking and suitable for sharing on social media.*
3. ***Adaptability for Dietary Restrictions:*** *Suggest options for making each dessert vegan, gluten-free, or catering to other dietary needs without compromising on flavor or texture.*
4. **Cultural Storytelling:** *Provide background or inspiration behind each dessert concept, tying it to Thai culture, history, or traditions, while explaining its Western twist.*
5. ***Feasibility for a Restaurant Setting:*** *Include considerations for preparation time, ingredient sourcing, and scalability, ensuring the desserts are practical for a restaurant kitchen.*
6. **Customer Appeal:** *Suggest names for each dessert that are memorable, culturally relevant, and enticing to customers."*

Example Output from ChatGPT

Thai Basil Lemon Tart with Coconut Meringue

- **Flavors and Ingredients:** A zesty lemon tart infused with Thai basil for a hint of peppery, herbal notes. The tart sits on a buttery, coconut-infused shortbread crust, topped with a fluffy coconut meringue.
- **Presentation and Plating:** Garnished with candied Thai basil leaves and toasted coconut shavings, served in a geometric tart shape for a modern aesthetic.
- **Adaptability:** Substitute aquafaba for the meringue and use a gluten-free crust for vegan and gluten-sensitive customers.
- **Cultural Storytelling:** Inspired by the balance of herbal and citrus flavors in Thai cooking, this dessert is a nod to the use of basil in both sweet and savory Thai dishes.
- **Feasibility:** Can be prepped in advance by pre-baking the tarts and assembling just before service. Ingredients like Thai basil and coconut are easily sourced.
- **Customer Appeal:** Name: *Coconut Basil Bliss Tart.* It appeals to adventurous eaters and those seeking unique flavor pairings.

Pandan Matcha Éclair

- **Flavors and Ingredients:** A French éclair filled with a creamy pandan-matcha custard, combining the grassy sweetness of pandan with the earthy bitterness of matcha. Topped with a glossy pandan glaze and a drizzle of white chocolate.
- **Presentation and Plating:** Served on a sleek black plate to emphasize the vibrant green glaze, with matcha powder dusted in

an artistic pattern around the éclair.

- **Adaptability:** Use gluten-free flour for the pâte à choux and plant-based cream for a vegan-friendly version.
- **Cultural Storytelling:** This dessert blends the iconic flavors of pandan, a staple in Southeast Asian desserts, with matcha, symbolizing the intersection of Thai and Japanese influences in contemporary cuisine.
- **Feasibility:** Requires skilled pastry preparation but can be batch-made and stored for efficiency during service.
- **Customer Appeal:** Name: *Pandan Matcha Dream.* The vibrant colors and unique fusion of flavors make it Instagram-worthy and enticing.

Tamarind Brownie Sundae with Coconut Ice Cream

- **Flavors and Ingredients:** A rich, gooey brownie infused with tangy tamarind puree for a sweet-sour twist, paired with a scoop of creamy coconut ice cream. Drizzled with salted caramel sauce and sprinkled with crushed peanuts.
- **Presentation and Plating:** Served in a mini cast-iron skillet with the brownie warm and the ice cream melting slightly over the top, creating a comforting and rustic presentation.
- **Adaptability:** Swap traditional flour for almond flour to make the brownie gluten-free, and use vegan coconut ice cream for a plant-based version.
- **Cultural Storytelling:** The tamarind adds a unique depth, inspired by its frequent use in Thai savory and sweet dishes, creating a surprising yet harmonious fusion with the rich brownie.
- **Feasibility:** Easy to prepare in large batches; brownies can be baked ahead of time, and ice cream and caramel are added à la minute.

- **Customer Appeal:** Name: *Tamarind Bliss Sundae.* The warm, gooey texture and innovative tamarind twist make it a crowd-pleaser.

Making Each Prompt Your Own: Customization Tips

- **Restaurant Branding & Style:** Share the vibe of your restaurant, whether it's casual, fine dining, or something else, to help ChatGPT craft desserts that align with your branding. For example, "We run a modern casual restaurant, and we want the desserts to be playful yet sophisticated." ChatGPT can adjust the complexity of the dishes and their presentation to match your concept.
- **Flavor Profiles & Customer Preferences:** Let ChatGPT know about any specific flavor profiles your customers enjoy. For instance, "Our guests love tropical fruits like mango and coconut, so we'd like to highlight these in our desserts." ChatGPT can tailor the fusion desserts to incorporate the flavors your customers gravitate toward, ensuring maximum appeal.
- **Highlight Specific Ingredients:** If you want to emphasize certain ingredients, such as locally sourced fruit or a signature product (like house-made ice cream), share that with ChatGPT. For example, "We want to feature locally grown mangoes in all of our desserts." ChatGPT can then suggest creative ways to incorporate these ingredients while maintaining a fusion of Thai and Western influences.
- **Dietary Needs & Inclusivity:** If you aim to cater to dietary restrictions, let ChatGPT know. For instance, "We want to offer a few vegan and gluten-free options without compromising flavor." ChatGPT will suggest substitutions and adjustments that ensure

the desserts are both inclusive and delicious for all guests.

- **Seasonal & Themed Menus:** If you're looking for seasonal desserts or options for special events, be specific about the theme. For example, "We're planning a summer menu with tropical flavors, and we'd like fusion desserts that incorporate coconut and mango." ChatGPT will craft seasonal desserts that are perfect for the time of year, highlighting fresh, in-season ingredients.
- **Presentation & Plating Styles:** If you have a particular style of plating in mind, share that with ChatGPT. For example, "We'd like the desserts to be Instagram-worthy, with elegant and minimalist plating." ChatGPT can then describe presentation ideas that align with your restaurant's aesthetics, whether that's sleek and modern or more rustic and traditional.
- **Kitchen Capacity & Skill Level:** Let ChatGPT know about your kitchen's capabilities. For example, "We have skilled pastry chefs, but we need desserts that can be prepared in advance and served quickly during service." Based on this, ChatGPT will propose desserts that are both practical for your kitchen and easy to execute during peak hours, while still being unique and exciting for customers.
- **Special Event Integration:** If you're planning for specific events (e.g., holiday specials, themed nights), include this in the prompt. For example, "We are hosting a Thai fusion dessert night and want creative ideas that reflect this theme." ChatGPT will suggest desserts tailored to the event, ensuring they tie into the theme and create an immersive experience for your guests.
- **Playful vs. Refined:** Let ChatGPT know the tone you want for your desserts—whether they should be fun and playful or refined and elegant. For example, "We want the desserts to have a fun, playful vibe, perfect for family diners." ChatGPT will generate options that suit this direction, ensuring the desserts align with

your desired atmosphere.

By sharing these key details, you can ensure that the fusion desserts suggested by ChatGPT will be perfectly tailored to your restaurant's concept, customer preferences, dietary needs, and kitchen capabilities, making them a seamless fit for your menu.

Conclusion

I hope you found value in exploring the expanded ChatGPT prompts in this **Bonus Chapter**. By now, you should have a better understanding of how small adjustments to these prompts can significantly improve their effectiveness, tailoring them to meet your restaurant's unique needs. With these insights, you're ready to start using the prompts that align with your business, helping you save time, improve efficiency, and streamline communication.

10

Frequently Asked Questions (FAQ)

1. How can I get the best results from ChatGPT for my restaurant?

Getting high-quality responses from ChatGPT depends on crafting clear and specific prompts. Always provide enough context, such as your restaurant type, target audience, or specific goals. Use the examples provided in this book as a foundation, and don't hesitate to refine and experiment with your prompts.

2. Will ChatGPT completely replace human input?

No. ChatGPT is a tool to assist and enhance your work, not replace it. It can automate repetitive tasks, generate ideas, and provide insights, but human creativity, decision-making, and personal connection remain essential in the restaurant business.

3. Can I use ChatGPT for tasks other than those mentioned in this book?

Absolutely! ChatGPT is a versatile tool that can adapt to a wide range of use cases. Beyond the examples in this book, you can experiment with it for employee performance reviews, brainstorming events, or even creating unique decor themes.

4. Is ChatGPT accurate when it comes to financial calculations?

While ChatGPT can help you analyze costs, create budgets, and suggest strategies, it's not a substitute for professional accounting software or advice. Always verify its outputs, especially when it comes to important financial decisions.

5. How much does ChatGPT cost?

ChatGPT offers free and subscription-based plans (such as ChatGPT Plus). The cost depends on your usage and the features you need. Visit OpenAI's website for up-to-date pricing information.

6. What are the limitations of ChatGPT?

ChatGPT is not perfect. It can sometimes generate responses that are overly generic, lack nuance, or are factually incorrect. Always review and refine its outputs, and combine its assistance with your industry expertise.

7. How can I ensure my staff and customers are comfortable with AI?

Transparency is key. Let your staff know that AI is being used to support their work, not replace them. Similarly, ensure that customer-facing communication retains a personal, authentic touch, even when using AI to assist.

11

Conclusion

The restaurant industry is fast-paced, competitive, and constantly evolving, requiring managers to juggle numerous responsibilities while finding creative ways to stand out. That's where ChatGPT steps in, offering a unique, powerful tool to help you streamline operations, engage with customers, and boost your restaurant's success.

Throughout this book, we've explored how ChatGPT can transform various aspects of restaurant management, from crafting compelling marketing campaigns to managing staff schedules, enhancing customer service, and analyzing financial data. With its ability to generate actionable insights and automate repetitive tasks, ChatGPT frees up your time, allowing you to focus on what truly matters—creating exceptional dining experiences for your customers.

But the real power of ChatGPT lies in its flexibility. It's not a one-size-fits-all solution; instead, it's a tool that adapts to your specific needs. Whether you're looking to develop creative menu ideas, respond to customer reviews, or create a budget for the next quarter, the prompts provided in this book are just the beginning. With experimentation and a touch of creativity, you can customize ChatGPT's capabilities to fit your restaurant's unique brand, cuisine, and customer base.

AI technology like ChatGPT is not just about convenience; it's about innovation. It's about exploring new possibilities for growth, building stronger customer relationships, and making smarter decisions. The restaurant industry is entering a new era, and those who embrace AI tools will find themselves better equipped to adapt, thrive, and lead.

We encourage you to take what you've learned in this book and dive in. Test the prompts, tweak them, and tailor them to your specific challenges. Explore how ChatGPT can help you save time, cut costs, and improve efficiency in ways that you never thought possible.

Ultimately, success in the restaurant industry is about passion, creativity, and connection. With ChatGPT as your ally, you'll not only have more time and resources to devote to these core elements—you'll also have the tools to elevate every aspect of your business.

The future of your restaurant is in your hands, and ChatGPT is here to help you make it extraordinary. Now, it's your turn to take the next step. Experiment, innovate, and see where this exciting technology can take you.

Here's to your success!

12

Thank You for Reading!

Thank you for taking the time to read *ChatGPT for Restaurants: 60+ Proven Prompts to Streamline Operations, Optimize Daily Workflow, and Accelerate Business Growth.* I hope this book brought you value, sparked new ideas, and inspired creative ways to enhance your restaurant management with ChatGPT.

I understand that not all of the prompts may apply directly to your specific restaurant or be relevant to every type of establishment. However, I truly hope you found some of the insights useful and that you enjoyed experimenting with the various ideas. Whether you're an AI expert or just starting to explore how technology can improve your operations, I hope you discovered something practical to apply.

If you enjoyed this book or found it valuable in any way, I'd be incredibly grateful if you could leave an honest review. Your feedback not only helps me grow as an author but also helps future readers understand the book's value and decide if it's the right fit for them. Your review means a lot and makes a real difference.

Thank you again for joining me on this journey. I wish you continued success, creativity, and innovation as you take your restaurant to new heights with ChatGPT!

Warm regards,

Brett Carroll

Made in the USA
Monee, IL
04 November 2026

67385898R00108